MY EXPERIENCE OF COVID- 19 BEFORE THE WORLD KNEW IT

ISBN: 979-8-9910186-3-0

EMAIL macro@rosanna.com

WEB SITE www.thehealthmode.com

Published by Rosanna.com
Book cover design by Rosanna Martella

TABLE OF CONTENTS

CHAPTER 1

MY EXPERIENCE OF COVID 19 BEFORE THE WORLD KNEW IT

I started to write this book in February 2022 because I wanted to tell everyone not to be afraid of the amount of evil that is around us, which is made of lies and deceptions. We need to be strong, no matter what we are exposed to. Share the truth, don't just hold back information for yourselves. Holding back out of fear and persecution creates more fear and persecution. Share the knowledge and share the truth.

In the beginning of June 2019, I took my granddaughter Francesca to her dancing rehearsal that was going to be held at the Catholic high school near where we live. We arrived early and were directed to the cafeteria where all the dancers were going to change into their costumes. The mothers helped all their children get ready with their costumes, fixed their hair and applied some makeup.

Francesca loved to dance and as soon as she was ready in her costume, she played and danced with the other girls until they were ready to go on stage in the theater.

The recital was going to be performed the following weekend.

The morning of the recital Francesca was not feeling good. She vomited and had a fever. She did not make it to the recital. I thought that she had caught some germ in that cafeteria.

It is strange but the year before (June 2018) Francesca had caught a cold in the same cafeteria, but that time it was not as bad as in 2019. She was able to perform at the recital that year.

I recalled that the very first time we entered the cafeteria of that Catholic high school I didn't like the energy of the place, plus they had all the ceiling fans running full strength which made the room very cold.
Now in 2023, my intuition tells me that the cafeteria was intentionally seeded with the corona virus from Fort Detrick with the help of some of the Vatican workers in 2018 and that this was only a light prelude. At that time Francesca had a runny nose with no fever though, but she was expelling a lot of mucus for a long time, no matter how much vitamin C I gave her. She was 4 years old at that time and never had a cold before in her life. Covid-19 was called 19 because that was the year they needed to unleash the full strength of the bio-weapon.

I took care of Francesca. Her fever was gone by the next day. She had a runny nose and she was tired the first and second week of June 2019.

I assumed that she had caught a cold like the year before. I continued to assist her for the entire week giving her fresh lotus root tea and soups to help her expel the mucus.
By the end of the week she had regained most of her strength even though she still had a runny nose and continued to take naps in the afternoons.

Eventually I learned from Dr. Mercola that children have immunity to Covid-19 and for this reason my 5 year old Francesca was able to bounce back faster than me, at age 71.

In the beginning of the third week of the month of June I began feeling sick. I thought that I had caught the cold from Francesca, just the way it had happened the year before.

Before that time I hadn't had a cold or flu since I started a macrobiotic lifestyle 30 years before. I had never before in my life experienced such an illness.

Every day there was a tremendous amount of mucus that I was coughing up. I had a very high fever and no appetite. Still I didn't go to a doctor, convinced that I could cure myself. I was taking two cups of fresh lotus root hot tea a day, because this natural remedy expels the mucus from the bronchial tubes and the lungs. After a few days I became so weak that I didn't have the strength to grate and squeeze the pulp of the lotus root any more. Francesca's father came every morning and did it for me.

I was sick for two weeks, from the middle of the third week in June through the first week in July. I was coughing copious amounts of mucus every day for more than two weeks. Strangely I was able to breathe through my nose at night and slept ok. It was definitely not a head cold. I had high fever for many days and then finally the fever lowered but it would not go away. I was still very weak, and I started to get worried. But still I didn't go to a doctor. I forced myself to eat mostly brown rice cooked in a pressure cooker with 10 cups of water and a small piece of Kombu sea vegetable. I would sprinkle sesame salt gomasio on my rice. At times I would eat some fresh steamed vegetables and miso soup. In two weeks I lost 17 lbs. of body weight. I am not a heavy person to begin with.

When the fever finally broke and the cough subsided, I still felt very weak until the end of July when I pushed to drive myself, my daughter and granddaughter to the Atlantic Ocean a few days a week, thinking that I needed Sunlight and the sea breeze to resurrect my body with a lot of vitamin D from the Sun. I had to lie down and take a nap every afternoon until the middle of September.

As time went by I still was convinced that I had a very bad flu/cold, until I came across a YouTube video in Italian language. This was during the time that in Italy people were more infected than other European nations and the majority of the people were dying mostly in the North of Italy. I was worried about my son who lives in Torino with his wife and their children.

The person that was interviewed in the video was a young military athlete who had gone to China in the month of October 2019. He said that the Chinese government had invited the Italian military athletes and other European countries for a military world game. He said that when they got there they were staying in a place near Wuhan.

After a few days there, because the place was dirty and they did not like the food prepared by the Chinese, they bought fresh food and cooked their own meals. Some of these strong, healthy military athletes got sick, including him and they went to see a doctor.

The doctor told them that there were many people around that area who had come down with the same type of illness that they had.

The young fellow proceeded to say that they had never gone down to the Wuhan market place the entire time they were there. He continued to say that he had a high fever and a very strong cough full of mucus and that he never, ever felt so sick in his entire life. When the fever got lower, it still didn't go away for a long time. He had no appetite and he felt very weak and tired.

At this point I woke up to the fact that what I had in June/July 2019 was Covid-19, because the symptoms this man was mentioning were the same as the symptoms I had in June. But how was that possible if I didn't travel to China and I got sick in the month of June, four months before he did? No one had mentioned Covid-19 the entire year of 2019.

Then I remembered that during the month of August I had heard somewhere in the media that a laboratory at Fort Detrick in Maryland had to be closed because they had an escape!!?!!! They closed it in August. This word "escape" remained ingrained in my mind because after hearing it I was saying to myself "what kind of escape was that?" If that was the escape of Covid-19… the China story did not hold well... We had the virus before the China "escape"…if it was an escape!

This was still brewing in my head a few months after, when the Pandemic was announced. By the month of December I started to connect the dots and if they did not say what type of escape this really was, then I believe it had to be some very evil program.

Maryland is only 2 to 3 hours drive from where I live. If someone in the lab had gotten infected and had been in contact with other family members and friends, other people could have gotten infected, and that would have had to take some time to get to us in New Jersey in the month of June. So the "escape bio weapon" had to have been going around for a good while. Why did the people in charge take all their time before deciding to close the lab finally in August 2019?

In June 2021 the Chinese newspaper Global Time urged the international community to look into "suspicious" activities at American biological research institutions, particularly the US Army Medical Research Institute of Infectious Disease at Fort Detrick, Maryland.
This news reinforced my understanding of how Francesca and I got infected in June 2019.

On another occasion I read that a Chinese defector claimed that the Military World Games had initiated Covid-19 outbreak in Wuhan, and it was intentional.

In my mind I began calling it the "Plandemic Industrial Complex."

By the time that the authorities announced the Pandemic, the infection was galloping in the world at large. I was certain now, my intuition was telling me, the whole thing was premeditated evil by the Deep State, Cabal, Illuminati, etc., who wanted to kill 90% of the world population.

I started thinking that all those infected military athletes returned home to Italy in October and unknowingly spread the disease through the population in Italy and Europe. It was clear to me that this was a planned program of the dark forces, to kill the population of the world. It looked like the beginning of Armageddon.

Why in Italy was the infection spreading so much with all kinds of people dying to the point that the authorities didn't know where to put the cadavers?

I had known for years that the Illuminati/Cabal/One World Order wanted to kill 90% of the world population and make the rest of us their slaves. So this could have been the preview!

Today, March 14, 2023, in Q The Storm Rider/Official page, I read:

The "Real Truth" Corona 19 virus was released in the U.S. in 2019 near FORT DETRICK military base… the Maryland NIH Fauci laboratory leak story even made it to the NEW YORK

TIMES in August 2019 (but was covered up quickly). DEADLY GERM RESEARCH IS SHUT DOWN AT ARMY LAB OVER SAFETY CONCERNS.

After CDC supposedly investigated the lab leak of the virus and closed the lab in 2019, people in the Maryland area had respiratory illnesses and infections and doctors reported the respiratory illness as strange and uncommon. The CDC then reopened the lab in 2020 as a COVID LAB (?) and CDC erased the lab leak story from the Archives.

There were two branches of the Lab in summer 2019 reported by MILITARY.COM. CDC inspection findings revealed more about the Fort Detrick Research suspension (after this article dropped, CDC scrubbed their Archives and the story was quickly covered up and [DS] MSM quickly dubbed the information of Q.ANON CONSPIRACY and the story was swept under the rug and whitewashed.

Scientists were working on the coronavirus at Fort DETRICK in March 2019 for NIH/ National Library of Medicine article "BROAD SPECTRUM CORONAVIRUS ANTIVIRAL DRUG DISCOVERY" by Sina Bavari from the division of Molecular and Translational Sciences, United States Army Medical Research Institute of Infectious Disease, Fort Detrick, MD, USA article by Sina BAVARI.

A few months later the lab had its 2nd leak and residents became ill with respiratory infection and sickness.

Behind the SCENES: XI IS COLLAPSING the CDC REGIME and arrested hundreds of officials, scientists, military commanders that have colluded with the CIA. OBAMA REGIMEN that also created the Covid-19 in Wuhan/Chinese military intelligence reports also state Fort Detrick as one of the beginning points and origins of BIO-WEAPONS Virus.

THE TRUTH is>> This is obviously a man made virus, it looks like being made a Frankenstein, engineered creation and that's the reason why there are no photos of the virus to this day…and the 5G also mimic the respiratory illness in the body and 5G was obviously a military weapon that was created by the U.S. DOD dating back to the 1970s.

So we have the gain of function Bioweapons, working alongside 5G military weapons and chemicals spraying frozen spike proteins (among other things) into our skies and letting us breathe the toxic Bio-weapons….

EVERYONE IS CONFUSED IF 5G OR CORONA IS REAL OR NOT. [It was designed this way to create confusion and fear, but also to infect a lot of people using all methods to create respiratory illness.] Whatever the matter of facts is IT'S ALL BEING EXPOSED.

I remember that during the time that the death toll in Italy was very high, there were people around that country saying that the Covid-19 and the Corona infections were caused by the 5G radiation.

Some years ago I came across this article in Italian. In a few words it described a truth that unfolds a difficult reality. Today I decided to translate in English, because from this point on I will report the horrible scenario of the reality I discovered through the many years of my research/study of those that some call the Deep State. (What you are going to read from now on until the end of the Matrix Chapter is very difficult to digest, so go slowly.)

"The cursed millennial global system of lies, which has deliberately been in place for millennia on this planet - laboratory - breeding, through the shameful and criminal tools of disinformation, mental manipulation and mass indoctrination, such as religion, mass media, history and science officers (trained servants and employees of the aforementioned system of lies), are giving in and we hope that sooner or later they will be destroyed and uprooted forever!"   ~~Unknown~~

As I investigated, the deeper I went in, the bigger the core grew of the betrayal imposed on human species. The anti-human agenda that formed this artificial program that is using chemicals GMO, electromagnetic, EMF radiation, political and religious manipulation is to raise pathological obedient slaves. All of this colludes with genetic experimentation by predators from beyond the stars in order to disconnect us from our original DNA blueprint.

When we ignore disharmony around us, evil doesn't stop, it gets stronger and continues to play ugly realities. We need to know what is going on beyond the scene. Instead of ignoring what is happening in our planet that is filled with horrors, we must be aware and expose the evil.

When we clean our home, we don't push the dirt under the rug, we dispose of it into the trash bin. The same is for the dirty horrible evil that goes on Earth, it needs to get exposed. We cannot ignore it. To turn our heads away from the evil horrors going on, and keep on partying, focusing on material gain, going shopping because it will make us "feel better" about ourselves, it doesn't liberate us from this immense EVIL.

Our Mother Earth is a conscious living planet. It has life force and will not allow low levels of human consciousness to continue to harm her or the rest of the awakened humanity. Instead of ignoring what happens around us, we can consciously surround it with LIGHT and ask SOURCE of all things to help us heal and get rid of all of this EVIL.

PTR tests don't diagnose infectiousness.
Inflation of death numbers,
Media propaganda, and rampant disinformation by governments,
Vicious cycles of lockdown Restrictions in movement and Travel
Arbitrary rules: curfews, temperature checks
Vaccine certificates for freedom
Loss of personal liberties
Loss of education
Loss of lives and livelihood

Fear, Fear, Fear of reinfection, fear of mutation and variants...continues to keep the fear going. Courage is not the absence of fear; courage is the triumph over fear. They, the puppeteers, fear those with knowledge and control those without it.

Which one are you?

What if I told you, your perception is a misconception because of media deception? There is NO climate crisis on earth. It's all a made up story. There is NO need to fear climate.

Earth is NOT at risk from humanity. Humanity is NOT at risk. It's all fabricated to add fear. They, the evil ones, create illusions to make us become their slaves.

December 5--2023

Hidden in the Covid-19 vaccines and the upgrade busters that some call it jobs are loaded with graphene nanobots oxide, mercury, spike protein and more toxins. These microscopic entities of nanotechnology can kill. They have the ability to get in all parts of your body. They navigate through the bloodstream and interact with the cell, tissues and even DNA.

Don't panic, there is the very first detoxification protocol for jobs side effects. The protocol was published in the US medical literature that gives people a chance to take this all into their own hands and uses natural substances to begin to help the body clear this very dangerous protein from the cell and tissues.

It involves 3 natural substances that are available over the counter in any type of online retail or natural food stores and some pharmacy. I also found it on Amazon.

They are Natto kinase, which is an enzyme derived from fermentation of Soy,

Bromelain which is an enzyme derived from the stems of pineapple

Curcumen, also called Turmeric a powder extracted from a ruth of a plant from India

The doses are :
Natto kinase 2000 unites 2 times x day
Bromelain 500 milligrams x day (it is a approved FDA drug)
Curcumen 500 milligram 3 times x day

For those people that had only one shot they must follow this protocol for 3 months.

For those people that have taken multiple shots they have to follow the protocol for one year.

Today January 30, 2023 https://benjaminfulford.net/

New peer-reviewed scientific studies have revealed what many of us knew from the beginning: 5G radiation is not only connected to the Covid-19 pandemic, it actually induces the body to create new viruses and illnesses, including coronaviruses.

And before the mainstream media gets hold of this study and convinces the masses that it is unimportant, you should know that these are peer-reviewed scientific studies published on the National Institute of Health website.
https://rumble.com/v274idk-u.s.-government-admits-5g-radiation-causes-covid-19-stunning-admission.html

This means not only the pharmaceutical companies but also the IT companies were involved in an attempt to murder a large percentage of the worlds' population.
This sort of information is coming out in the mainstream now because the Khazarian Mafia (KM) are losing control of the Fortune 500 companies.

However, the story runs much deeper than that. The reason Western drug companies are forced to create diseases and then sell the cures is because they are forbidden from producing drugs that make us smarter, happier, stronger etc. A pharmaceutical company executive once told me nothing would be easier than to make drugs that would increase our IQs but that they were forbidden from doing so. He also said making new recreational drugs that were not harmful "would be incredibly easy." The reason athletes are banned from "doping" to make themselves stronger, faster etc. is the same.

It is because the Octagon group at the top of the KM is deliberately holding us back as a species. They forbid drugs and genetic changes that would increase IQ etc. They also deliberately keep our lifespans artificially short so we do not have enough time to figure out the control matrix we are subject to. In other words, defeating the KM will be much bigger than the fall of the Soviet Union. It will change everything. Computers are said to double in processing power every two years. Humans could do the same if they wished to and were allowed to.

———◦⫷◉⫸◦———

Dr Mercola from a 2022 newsletter

Medical technology is an economic ideology built around totalitarian rule by unelected leaders that got its start in the 1930s, when scientists and engineers got together to solve the nation's economic problems.
Evidence of technocratic rule was also evident during the pandemic. The censoring and manipulation of medical technocracy has lied to us about the risk of death from Covid-19. Based on death per capita, the global average death rate for Covid-19 is 0.009%. The average person's chance of surviving this disease is 99.991%.

From Dr Mercola March 18-2023

There is a surge in deaths and disability that has occurred since the COVID-19 busters shots campaign rolled out.

Former BlackRock analyst and fund manager Edward Dowd is one of the brave few who have been trying to get the word out about dangers of COVID-19 shots. In his book, "Cause Unknown: The Epidemic of Sudden Deaths in 2021 and 2022," — his information is finally getting mainstream media attention.
Group life policyholders, who are typically healthier than the general population, experienced mortality spikes of 40% in 2021

Disability numbers among the workforce reached a high of 33.2 million in September 2022, with numbers still trending up — a highly unusual increase

In an interview with Tucker Carlson, Edward Dowd explains that media outlets like Yahoo have picked up on the undeniable increase in deaths among young, healthy adults. However, they're quick to state that such deaths are not due to COVID-19 shots. But Dowd isn't deterred. As A Midwestern Doctor noted on Substack:

"Ed Dowd has focused on utilizing a narrower set of evidence and tying it to one of the most persuasive arguments currently available for shifting the narrative. A statistically impossible spike in sudden deaths has occurred in the healthiest segment of the population and has happened in tandem with a spike in disability (this is why we are now having labor shortages)."

Down believes there's enough alarming data to warrant the Covid-19 shot program being stopped immediately, as the death and disability from the shots could easily exceed from Covid-19.
Central Bank, pharmaceutical companies, Big Tech and the media all benefited from the pandemic and have an interest in covering up what Dowd describes as a " large global murder scene".

⸺◦⫷◆⫸◦⸺

Today March 29- 2023 on Children health Defence News and Views, I read this:

In the U.S., COVID-19 vaccines injured 6.6 million people, disabled 1.36 million people, caused more than 300,000 excess deaths and cost the economy an estimated $147 billion in damage — in 2022 alone — according to a new analysis by Humanity Projects, a wing of Portugal-based research firm Phinance Technologies.

The researchers behind "The Vaccine Damage Project," released this month, said they sought to "estimate the human cost," including "deaths caused or hastened by the vaccines," as well as "the impact on the overall economy of each aspect of the vaccine damage."

Phinance Technologies was founded by former BlackRock portfolio manager Edward Dowd, along with Yuri Nunes, Ph.D., and Carlos Alegria, Ph.D.

Dowd, who came out as a whistleblower against the COVID-19 shots and Big Pharma corruption, is the author of " 'Cause Unknown': The Epidemic of Sudden Deaths in 2021 and 2022."

https://childrenshealthdefense.org/defender/covid-vaccine-injury-deaths-economic-damage/?utm_source=luminate&utm_medium=email&utm_campaign=defender&utm_id=20230329

━━◄◊►━━

Today December 9--2023 Dr. Mercola wrote in his Newsletter

In an October 2023 lecture, David E. Martin, PH.D., detailed how we can know that SARS-CoV-2 is a man made bio weapon that has been in the Works for 58 years.

The virus called "coronavirus" was first described in 1965. Two years later, the U.S. and U.K. launched an exchange program where healthy British military personnel were infected with coronavirus pathogens from the U.S. as part of the U.S. biological weapons program.

In 1992, Ralph Baric at University of North Carolina, Chapel Hill, took a pathogen that used to infect the gut and lungs and altered it with a chimera to make it infect the heart, causing cardiomyopathy. This research was part of efforts to produce an HIV vaccine.

In November 2000, Pfizer patented its first spike protein vaccine. Between 2000 and 2019, vaccine trials using this technology proved it was lethal, yet in the summer of 2020, the clinical trials for the SARS-CoV-2 shots went straight into human trials.

mRNA spike protein was publicly described as a bioweapon 18 years ago. In 2005, at a conference hosted by DARPA and The Mitre Corporation, the mRNAspike protein was hailed as a "biological warfare-enabling technology", i.e., a biological warfare agent.

━━◄◊►━━

Emotions: We need to go within our body and stop the fear. There, inside of us there is nothing to be afraid of. Our emotions are something to be managed. We must not avoid our feelings! When you ignore them they go underground and come out in more destructive ways. We need to see emotions as a call to action and realize that we are having these emotions as our friends because something needs to change. They are the things that will make us act to improve ourselves.

Without emotions we are like a tree without leaves, a flower that never blooms, a bird without wings, we will be disabled. Emotions are not a burden. If you care only about your safety we will all be destroyed. We need to unite and work together. When we unite with other like minded people that don't want to see any of us die, we learn to face our fears and work together to realize how to win the time for survival of all. Emotions are very important to us, they contribute to empathy, which the controllers don't have.

Awakening
Since the beginning of our self awareness, at the start of being incarnated in a body, we have looked in this world for clues to understand the meaning of our existence. Why is there so much good and so much evil on Earth?
Hoping to find an answer that brings an understanding, I started to investigate the stories of reincarnations. Then I discovered that we carry within us, all the lives lived by our Soul.
In the Akashic Records of our heart chakra, we carry all the incarnation stories which are never lost, they are saved within the Akashic Records of our Soul.
The Living God is within us in every fiber of our being.
As we awaken our consciousness we realize that everything is God, we are created in the vibration of the Creator of All Things, and our Soul has an origin, we are multidimensional Sacred Souls.

This below is a little piece from the EMERALD TABLETS written around 36,000 B.C. by Thoth the Atlantean, Priest-King. This manuscript precedes any Egyptian writings.

Tablet # 7

"Wisdom is everything; do not be silent when evil is named so the truth is like the light of the Sun to shine on everything.

Whoever tramples on the law will be punished, because only through the law is the freedom of men obtained. Do not cause fear because fear is slavery, a chain that binds hearing to darkness. Follow your heart throughout life. Do more than you are commanded to do."

Tablet # 10

"Know that knowledge is reached only with practice and wisdom is created only with knowledge and in this world the cycles are only created by the law, they are tools for the conquest of knowledge. Because the plan of the law is the source of everything".

Let's get involved in rituals that connect us to the DIVINE REALMS. Chanting, mantra, praying in our own way, with our own words to bring peace in the world, fill all with hope and connect daily with the DIVINE.

If we don't stop the evil, but we let it expand, it will be the end of our planet and all of us.

We have HOPE even if…

The night seems darker and darker before dawn…. DIVINE LIGHT is pouring on to us, more than ever before.

CHAPTER 2

PHARMACEUTICALS

Big Pharmaceutical spends $5 billion per year lobbying Congress.

And this is how they make the way for passing laws that they impose on the unaware citizens.

In a research study made more than ten years ago, on 1,200 patients, they analyzed the toxicity in the blood and found 20 toxins in average people. Ten years later the number went up to 500 toxins. Now we must always detox our bodies. Detoxing is no longer a hobby; it is a survival strategy for everyone. An intoxicated population that cannot think clearly is much easier to manipulate and control. We are in a fight for our lives, right here and right now. Vaccines are one other way to keep us dumbed down, weak, sick and some dead.

The vaccines that Big Pharma uses are infested with substances that shouldn't be there, not even in microscopic amounts, because they are potentially harmful for people's health and wellness.

Over the year 2021 the US government spent $1 billion of US taxpayers' money to advertise Covid Jabs which is the most dangerous and least proven drug ever marketed. While simultaneously they are calling for censorship of anyone who dared to address the risk of this novel treatment upgrade.

I read on Dr Mercola.com:

"By law, drug ads must not be false or misleading, and must present a "fair balance" of information describing both the risks and benefits of a drug, must include facts that are "material" to the product's advertised uses, and must include a "brief summary" that mentions every risk described in the product's labeling. Few if any ads for the Covid jab have fulfilled these requirements."

⬤⊱⬤⊰⬤

Also vaccines for diphtheria, tetanus, whooping cough, hepatitis B and others are having their analysis performed by government officials that are not disclosing the truth because the officials are corrupt and they do not protect the health of the citizens. The political system is sold to the pharmaceuticals.

The analysis of private labs has found a quantity of fetal DNA and antibiotics plus fungicide, herbicides and glyphosate that shouldn't be in there.

The private analysis results were revealing terrible things and the results were published on scientific international papers, and also were pinned up at the upper offices of the European Union, police stations and other competent agencies.

Then there are all the highly toxic elements of nanoparticles. These are the Geoengineering mixes that are raining down from the airplanes in the sky above us every day.

Planes are spraying our sky with highly toxic Aluminum, Glyphosate, Magnesium, Strontium Lithium, Manganese, Barium and more stratospheric Aerosol injections. We breathe this every day with terrible health consequences, Mental and Physical. (This is something that President Kennedy was speaking about before he was killed.)

If this is not enough they are injecting Aluminum and Mercury in adults and children's vaccines.

I remembered a part of a speech that

President J. F. Kennedy had made, few days before he was assassinated and now it was much clearer in my mind:

"We are at the tip of the transhumanism agenda replacement program (de-population). EMF mind control weapons are used on the population, chemicals, nano fibers and viruses are sprayed in our sky so we inhale them. This nano brain intraintainment is used to make brain-washed human robots that work and obey to any type of command."

Transhumanism, this word that President Kennedy mentioned in his speech is about replacing the human original DNA blueprint with the use of synthetic technologies, and artificial intelligence programming to destroy human's spiritual functions.
The evil controllers have been playing God with our physical body.

Their intentions are to build a human being that they can control completely. They can accomplish that by destroying our emotions. Emotions are very important to us, they contribute to empathy, which the controllers don't have.

President Kennedy was talking about "chemicals, nano fibers and viruses that are sprayed in our sky so we inhale it." He knew all of this way back then! He knew what was going on.

Is this speech from President Kennedy very real now? He had the intention to fight it!

We have been slowly poisoned every day a little bit more for hundreds of years. We are now in a fight for life right here right now. Humanity has to wake up to the evil going on.

In 2020 physical implants were administered to the population through vaccination programs by the Cabal/WHO/Pharma. That program has now been removed by the résistance of humans by a specific technology that can be operated from a distance.

Plasma implants that suppress higher knowledge have not been fully removed yet, but were made less effective in their functioning.

There has been a technology for some time now that is able to duplicate a human being in 5 months. These so-called Clones are programmed with memories, mannerisms, and speech patterns of the original human. Clones don't look exactly like the original because they have not lived the same long life of the original.

The Cabal/Deep State has been replacing VIPs as they are arrested, died, or have been killed. Just to mention, Hilary Clinton, Obama, Soros, Biden, are a few that I suspect have been replaced with clones.

The movie Avatar gives us a little idea how a clone is made.

Many Black operation programs and human cloning laboratories that were operated by the Cabal/Deep State have been cleared and shut down by the military resistance, but some are still working in secret places as there are still clones around. The life span of a clone is very limited.

There is a manipulation of the truth going on here. The puppeteers fire garbage from the news screens and Social media, because many of their puppets are clones or brainwashed individuals, which are used by those who intend to break down the fabric of our society.

As each lie unfolds before us we must remember that every lie has an opposite and we have the power to discover the truth on the other side. Everything that is illusory has no solid moral foundation, is based on lies and distortions.

There is a Law in the Universe that all the creatures have to abide by. Because of this Law, the Dark Forces have to ask our permission if we agree or not to what they want to do to us. So they put their story in Movies and other ways.

Most if not all scary movies are made for that reason. It is to show us what they are really up to; even though people don't realize it and believe that they are non real stories in scary movies, but are make-believe and entertaining. That way the evil ones continue to do their dirty work undisturbed, because we are condescending without really knowing it.

Let's lift our consciousness to truth! Let's move forward and leave darkness behind. Let's remember that we are all ONE, and we are more than them.

Today May 20, 2022 Dr. Mercola newsletter

Government Scientists Secretly Paid Off While Hiding Data
According to government watchdog Open the Books, the National Institute of Health and hundreds of individual scientists received an estimated $350 millions in undisclosed royalties from third parties, primarily drug companies, in the decades between 2010 and 2020.

Between 2010 and 2014, National Cancer Institute employees received nearly $113 million. The National institute of Allergy and Infectious Disease (NIAID) and its leadership received more than $9.3 million.

Federal agencies are increasingly refusing to comply with Freedom of Information Act requests, thereby forcing legal action. This is an obnoxious waste of taxpayer money as, by law, they're required to release information.

Forced FOIA disclosures have shown the NIH lied about not funding research in China, and allowed the EcoHealth Alliance-- whom they're supposed to regulate-- to write its own reporting rules. NIH has also been caught redacting information under false pretenses.

Members of U.S. Congress are calling for an investigation into the EcoHealth Alliance, to determine the true scope of its cover-up.

House investigators have found EcoHealth hid more data previously known, including a death rate of 75% in humanized mice infected with coronavirus.

Dr Mercola has explained since 2020, asymptomatic spread of Covid-19 is likely to be rare as to be non-existent.

It was a lie perpetuated to drive up fear and prop-up rising "case" rates that didn't exist. It's basic virology that you cannot transmit a virus unless you have a 'hot' infection, and if you have an active, transmissible infection you have symptoms. The symptoms are a sign that your body's defenses are kicking in to rid itself of the live virus.
NO SYMPTOMS, NO TRANSMISSION.

October 24, 2022, from Dr Mercola Newsletter

The Federal Aviation Administration (FAA) requires first-class airline pilots to receive an electrocardiogram (EKG) starting at age 35, and continuing annually after age 40.
EKGs record the heart's electrical activity to provide a measure of heart health and certain parameters must be met in order for pilots to be deemed fit to fly.

The FAA changed the EKG requirements necessary for pilots to fly — but not to make them safer. With no public announcement or explanation, the agency expanded the allowable range for the PR interval, a measure of heart function.

Widening this parameter means those with potential heart damage, disease or injuries are now allowed to fly commercial aircraft, potentially putting passengers at risk, should they suffer a heart attack or other event while in the air. Why would the FAA make such a drastic and risky move without informing the public?

———◦<⧫◦⧫⧫◦>◦———

Since the beginning of 2020 I have been saying that we are into World War Three, (WWIII) but this time the war is not only made with bombs and bullets, for the most part is made with needles and vaccinations.

March 18- 2023 on Dr Mercola Newsletter states the following:

Former BlackRock fund manager Edward Dowd is bringing attention to the surge in deaths and disability that has occurred since the COVID-19 shot campaign rolled out.
Group life policyholders, who are typically healthier than the general population, experienced mortality spikes of 40% in 2021.
Disabled numbers among the workplace reached a high of 33.2 million in September 2022, with numbers still trending up- a high unusual increase.
Central Banks, pharmaceutical companies, Big Tech and the media all benefited from the pandemic and have an interest in covering up what Dowd describes as a "large global murder scene".
Dowd believes there's enough alarming data to warrant the COVID-19 shot program being stopped immediately, as the death and disability from the shots could easily exceed that from COVID-19.

"Ed Dowd has focused on utilizing a narrower set of evidence and tying it to one of the most persuasive arguments currently available for shifting the narrative. A statistically impossible spike in sudden deaths has occurred in the healthiest segment of the population and has happened in tandem with a spike in disability (this is why we are now having labor shortages)."

Dowd is intent on bringing global attention to this surge in deaths and disability that has occurred since the COVID-19 shot campaign rolled out, and he's not willing to let anyone, or any entity, stop him. "We have the data. We have the evidence," he says, "and there's a large global murder scene that just occurred."

⸺•⋖⧓⧫⧓⋗•⸺

On March-2023 a friend sent to me an email with this:

My recent conversation with Dr. Bryan Ardis left me speechless…

"Do you Know that, right now, there are at least two snake villages in China that Harvest millions and millions of snake eggs for the vaccine manufacturing world?"

"Did you know they're harvesting vaccines and injecting children with snake eggs? (And it's been happening for years!)

I couldn't believe my ears, I knew for years that most of the world's vaccines have been produced in China…And that vaccine materials were grown inside chicken eggs. But, what I didn't know was that China secretly created snake villages where they could harvest millions of low cost snake eggs to produce vaccines, cheaper and faster.
"The title of the article is "Like Venom Coursing through the Body.""Researchers discovered the biomarker indicating mortality in COVID-19 patients and it's an enzyme found in rattlesnake venom".

DR, Ardis' next comment sent chills down my spine:

"If they are harvesting vaccines from snake eggs, that means the proteins of reptiles will be in your vaccines".

He told me that the University of Arizona published research in the summer of 2021 where they evaluated 300 COVID patients who had died in Hospitals.

"The title of the article is "Like Venom Coursing through the Body.""Researchers discovered the biomarker indicating mortality in COVID-19 patients and it's an enzyme found in rattlesnake venom. That's what they published".

"This substance is called phospholipase A2".

"And they published that every medical doctor around the world should be looking for phospholipase A2 as a biomarker to look in the blood of the patients in your hospitals, if the biomarker is going up, the patient's is going to die because this enzyme found in rattlesnake venom destroys multiple organs at once, and will develop organs failure and the person die".

Now the phospholipase is 20 times higher in COVID-19 patients than any other group of individuals that they'd ever seen with any chronic illnesses outside of Covid. So those with autoimmune disease, congestive heart failure, diabetes, and in the elderly, they sometimes see low levels of phospholipase A2, but they'd never seen it ever recorded in a human 20 times higher like the ones in the COVID patients.

Given what we've learned about COVID bioweapons and the global Elites" population control agenda, Dr Ardis' next questions made perfect sense:

"How do you know there's no snake-based phospholipase A2 protein that's being extracted from these snake eggs in their vaccine manufacturing and they're injecting them inside of you? You're developing disease, chronic illnesses, and they're finding low levels of phospholipase A2. Now do you know weren't introduced inside of you when you received your vaccine?

"This would explain a whole bunch of the chronic illnesses that individuals are living with. If they have figured out how to get snake protein into humans, they have created a new narrative that there's this new enzyme.

"How was venom, a venom component called phospholipase A2, which was first ever discovered in a snake? How is it now in low levels of critically ill people?"

Dr Ardis went on to talk about Stew Peters' incredible documentary called:

"Died Suddenly"

"It really starts with discussing these blood clot formations that are being found in people who are dying after being vaccinated. These blood clots are a side effect of the mRNA injections.It is also a component of venom found from multiple snakes in COVID-19 in Italy. They're called prothrombin activator V proteins. Prothrombin activator, pro coagulation factor V, which is a component of the blood clotting cascade triggered by snake venoms, and it's the only component that has any impact on blood.

It causes blood clotting.

Now snakes in the world have 2 primary blood side effects. When you get bitten by a viper, get venom in your body of any kind, there are only 2 actions, is either going to clot your blood or going to thin your blood. In the COVID-19 patients they only found the venom component that causes blood clotting. And it's called a procoagulation thrombin activator. So they isolated these components that cause blood clotting. Do you know this procoagulation Factor V isolated in these snake venoms?"

"I already have the research studies. They confirmed that this component is identical and does not need the mammals blood clotting factor to make the blood clots, it does it all by itself in the human body. So as soon as it goes into the human body, it starts the cascade to make red blood cells stick together, called Rouleaux formation."

It's clear that the Global Government has been using vaccines to spread chronic disease and create infertility for years. And that they crafted the COVID death jabs to speed up their nefarious depopulation agenda. In fact, top bioweapon expert, Dr. David E. Martin, who has researched many of the 4,000 coronavirus patients, told me that there have been spike protein depopulation patents since 2021.

Sadly, billions of people around the world were coerced into getting the deadly COVID bioweapon "vaccine"... Without realizing the Truth until AFTER they had already been Jabbed. That's because the Global Government, the Elites, Big Tech, and the Big Pharma-funded Corporate Media... Have been censoring, conceling. and deplattaforming everyone who dared to expose their lies..."

But no one should despair. We are blessed to be working with Dr. Ardis and hundreds of other brave experts who are risking their own careers to expose the TRUTH, and they're successfully treating their own patients with Healing protocols that are reversing the damage from the venomous spike protein.

So even though I KNOW how bad things are…I'm embracing Hope and HEALING.

'Your body was designed to stay well. You hold in your hands the power to take control of your health. Never let anyone take your right to health away.'

~ Dr. Mercola

'If the state can tag, track down and force individuals to be injected with biologicals of known and unknown toxicity today, then there will be no limit on which individual freedoms the state can take away in the name of the greater good tomorrow.'

~ Barbara Loe Fisher

Government, Big Pharma, and their lapdog MSM, were complicit in violating all 10 points of the Nuremberg Code with their Covid vaccine campaign.

1. Were the public given "voluntary informed consent"?

-No -We were lied to, and told everything was 100% safe.

2. Was the result good for society?

-No -It did NOT stop transmission and has proven to be harmful to many who were at little to no risk from Covid.

3. Were there prior experimentations on animals that "justify the performance of the experiment"?

-No -Animal experiments were conducted, but the results and validity of the data are in question due to the lack of data, and lack of safety/efficacy in humans.

4. Did they "avoid all unnecessary physical or mental injury"?

-No -We were subject to 24/7 propaganda, brainwashing, and coercion from employers, media, and government entities. Not to mention the negative vaccine side effects such as Pericarditis and Myocarditis.

5. Did they stop possible "lethal or disabling procedures"?

-No -Government health agencies continued to push the experimental shots despite the debilitating side effects and potential death. They went great lengths to cover up VAERS and all talk of vaccine injuries on social media.

6. Did the "degree of risk outweigh the benefits"?

-No ❌ -99% of people were not at risk from SARS-CoV-2, and the experimental mRNA shots did not prevent transmission.

7. Were proper preparations and facilities prepared to prevent "remote possibilities of injury, disability, or death"?

-No ❌ -Subjects were brainwashed and largely unaware they were being experimented on and did not know they were at risk of injury, disability, and potentially death.

8. Were the experimentations conducted by only the "scientifically qualified"?

-No ❌ -Scientists who spoke out about the mRNA shots were silenced, censored, and intimidated by government agencies. Pharma propagandists coerced lower level public health workers to administer shots that they did not know the true data about. The doctors didn't KNOW the vaccines were safe, they were told they were safe. They were wrong.

9. Can participants "freely end the experiment"?

-No ❌ -Most of them don't know they are being experimented on, and even if they do, permanent alterations have been made to their bodies via mRNA technology that cannot be undone as far as we know.

10. Did they stop the experiment when it "proved to be dangerous"?

-No ❌ -Not only did they not stop, they kept going, doubled down, and abused government emergency powers to silence and censor US citizens who spoke about the real dangers of the mRNA vaccines on social media.

This may not officially be accepted as law, but the actions of our government violate all medical ethical standards and infringe upon the basic human rights recognized by every sovereign nation on the planet.

We will have Justice.

CHAPTER 3

THE DEEP STATE, CABAL, WHO TREATY OF CENSORSHIP

I had known that there was The Agenda 21 made by the Deep State, Luciferian Cabal, and the New World Order (NWO) that was preparing depopulation of the world.

I had read some papers about AGENDA 21 around the time the military/resistance/white hats had asked Trump to attempt to become president.

This new strange president started to fight the Deep State from the inside of the corrupted infiltrated government from the first day he got elected.

Shortly after Trump's election, the Luciferian Cabal moved up the date of the Agenda from 2021 to Agenda 2030. Most probably because Trump was starting to fight the NWO with the help of the white hats military, from inside of the corrupt government and they felt threatened, so they gave themselves more time to prepare to accomplish their dirty deeds.

I recently read this below on Twitter and I transcribe it here in my own words:

"The hatred of those who speak ill of Trump is not based simply on the fact that he governed well. The fact is that he embarrassed them by making his job look easy! He made deals in minutes that those professional clowns could never have done in years. So they swore never again to let a man like that get near their power levels!"

This video link will show to you what Trump was doing during his presidency to unite other countries to fight the evil of the Cabal, deep State, Khazarian Mafia etc.

https://bestnewshere.com/the-trump-presidency-and-the-the-global-us-military-operation-to-drain-the-swamp-video/

What is going on now in 2022, is not just about vaccines, masks and Covid fear. Its ideology, it is a fight about whether the individuals are going to have control over their life, what they eat, the air they breathe, the water they drink and what they put on their body, and to have liberty of speech; because the puppeteers don't want us healthy, they want us submissive to their will like real wooden puppets.

I read this somewhere and it stuck in my mind:
"For the fear of dying we have stopped living."

For those that live in fear, maybe reading the book by Dr. Mark McDonald will help overcome the fear.

In his book "United States of Fear," psychiatrist Mark McDonald diagnoses the U.S. as suffering from mass delusional psychosis, driven by an irrational fear of what is now a rather innocuous virus.

In his newsletter Dr Mercola stated this below:

A treaty that would create Global Censorship of health information.

The World Health Organization (WHO) cannot be allowed to control the world's health agenda, nor enforce biosurveillance, while it receives funding from public sources belonging to the people. It is caught in a perpetual conflict of interest because it also receives substantial funding from private interests that use their contribution to influence and profit from WHO decisions and mandates.

For example the Gates Foundation and Gates-Funded Gavi vaccines promotion alliance, contribute over $1 billion a year.

We are not dealing with a democratic situation or body that serves the needs of the people. This is not an organization that world countries should entrust their health care to. This is One Health, One World Order- One organization, and one size fits all. (No one could complain if things go wrong, because of the diplomatic immunity they acquire).

The globalist Cabal wants to monopolize the health system on a worldwide scale with an international pandemic treaty. The negotiations for this treaty began March 3, 2022, and are expected to become a reality by 2024 as reported by Pulse in a YouTube video.
The Pandemic treaty is a direct threat to a nation's sovereignty to make decisions for itself and its Citizens. The beast wants to implement it throughout the governments of the world.

We need to believe in a community focused, ethical, decentralized as a way to approach health, that leaves this power in the hands of people to make decisions that are best for themselves. People have an innate wisdom and ability to decide for the best, so now from Australia to Zimbabwe people are standing against this proposal of a pandemic treaty and have decided that we must stop the WHO. There is a massive violation of human rights in the World Health Organization.

We must start to educate ourselves and:
1) Raise awareness about the implication of the proposed global pandemic agreement;
2) Call out the National campaigns that protect natural law and democratic constitutions;
3) Join credible society coalitions such as the World Council for Health. In the website worldcounselforhealth.org there is good information

---

In April 3-6, 2023 The World Health Organization (WHO) opened their annual meeting which is the World Health Assembly

"THIS IS A ANTI-HUMAN-AGENDA"

The World Health Assembly (PANDEMIC treaty) The negotiation body for the treaty is meeting.

How One Health efforts point to a deliberation scheme with host Meryl Nass, M.D. and James Colbett. They expose the corruption in the WHO and pose their perspective on how Big Oil's ties to climate affairs is worth considering. The Intergovernmental Negotiations Body for the treaty is meeting this week. April 3-6 -2023

They will have an open session at this meeting. They have an all series of meetings. They have a schedule to finish by May, but we don't know when they will finish but the probability will finish in May 2024. It is difficult to know when they are going to end because they are working as quickly as possible, almost all happen beyond close doors.

It is very hard to find out as much as we can of this process and get everybody ready, but most things are happening in the shade.

Last year in 2022 only 25 or 30% of the Nations of the world where there and the person with the gavel said: does anyone have any comments and concerns? Nobody said anything and in a moment he banged his gavel out and he said: "it is a done deal, we have consensus ". And that's how the amendments were brought forward last year.

We assume a similar process will go on this year, and there is always the issue that we don't have to comply. Your Nation doesn't have to comply; nobody has to comply if they don't want to.

James Corbet talked about the live webcast of the meeting of the Intergovernmental Negotiating Body.

This is the IBM that we are talking about for months now, and the last session token from the 3-to6 of April and on their site they have the provisional agenda, the draft program of work and a progress report of the IBM, they have the opening session and the closing session, everything else is taking places in close doors!!?

On their PROGRESS of the INTERGOVERNMENTAL NEGOTIATING BODY to draft and NEGOTIATE a WHO convention agreement or other international instruments on pandemic prevention, preparedness and response (INB) to the Seventy-Sixth World Health Assembly.

What was very interesting to see in their report was:
So this is the progress report for this particular session 3-6 April 2023 . But their progress report was drafted on the 28 of March 2023 and in point 12 it says:
At the fifth meeting of the INB further considered the zero draft WHO Ca+ and agreed on a process forward. Sothie was drafted on March 28 but they are talking about in April as if it is already done.

In the final point 13 we read: In the period leading up to the seventy-seven WHO assembly in March 2024, the INB will, pursuant to its agreed timeline and deliverables, hold 4 additional sessions (including " two weeks marathon sessions) in first quarter of 2024, as well as two additional sessions of the drafting group in 2023. The IBN may , if it finds appropriate, supplement in working sessions, in order to meet the ambitious deadline established by the Health Assembly for the INB's work.

In the link there is much more information.

https://live.childrenshealthdefense.org/chd-tv/shows/good-morning-chd/antihuman-agenda-with-james-corbett/

— ◦ ⊰◈⊱ ◦ —

In April 13, 2022 I read this in Dr Mercola newsletter:

"The price we'll pay for ignoring truth will be far more severe, as it'll cost us literally everything — our financial wealth, our material possessions, our health and bodily autonomy, our freedom and any possibility of pursuing happiness on our own terms."

— ◦ ⊰◈⊱ ◦ —

In a world full of chatter, listen to your own truth within.

We will prevail! We are stronger than we know! For we are LIGHT itself and we will make through this maze of deceit and lies. We are the forerunners of a new paradigm who will set the precedents of a way of living upon which the ground rules of a new peaceful happy World will be based on!

We are the children of LIGHT, we are very powerful! Let's find moments of laughter and find more moments to be thankful! This will raise our frequency.

CHAPTER 4

MY STUDY OF THE BOOK OF REVELATION: AND WHEN I BEGAN TO SEE THE DECEPTION

In the 1980s, I studied the Book of Revelations with a very enlightened teacher/ priest, Padre Quintin Walsh. He had been a professor of Eschatology in the catholic high school around my area for more than 20 years. One day a new Bishop was appointed in charge, and this newcomer got rid of that type of teaching in the Catholic high schools where Padre Walsh was teaching..

As a result, Father Walsh got very sick after they dismissed him. When he finally regained his strength, was sent to the Catholic Church near my home.

I used to go to church in those days, and when I listened to the first sermon of Father Walsh, I was floored. I had never been present to hear a sermon expressed so magnificently, in my entire life. I discovered that day that other people had asked him to teach the Bible to them and I immediately joined the group.

He taught us parts of the Bible from the Old Testament and parts from the New Testament including the book of Revelation. I followed him for more than two years even when he was transferred to another parish farther away from my home. The second year I studied the book of revelations again. It is a very difficult book!

This was during the time I was taking the anti-seizure medicine and the side effects were starting to give me added problems. It is explained in my previous book Healing Epilepsy Naturally.

As I learned, through the study of the book of Revelations, I didn't realize the fact that we were getting closer to the end of time the way we knew it. In those days I didn't think that it was going to unfold during my lifetime. But I was wrong! As I studied, the veil over my eyes opened up little at a time, revealing a difficult truth to swallow.

<hr/>

I grew up in Italy. My father was in the military. The day he graduated from the military academy, World War II started and he was sent immediately to war. My father served 7 years of war in Yugoslavia and Albania. He retired as a police chief in 1963 at the age of 45, I was 16 years old. When I asked him why he retired, he told me that he had given a lot to his country during the seven years of war and even after as a military officer and policeman. Suddenly he was transferred to Pagliarelle di Palermo in Sicily. He told me that he had no intention of getting killed by the Sicilian Mafia. So he resigned the same day that he got to Palermo and came home, to care for his family.

I became aware of what Martial law looked like, when I was very young. I learned how to research things that spike my interest.

I recognized how the Bush administration was attempting to take over America. When Bush Jr. arrived in power and began to talk about the "One World Order " it made me very uncomfortable. I witnessed Martial law with my own eyes in those ugly days, when the September 11 disaster happened, I clearly understood…A couple of airplanes wouldn't have had the power to make those three buildings collapse into their foundations!??

I studied and worked as an architectural designer, and I realized that the buildings fell because a planned demolition had been set up. After the authorities got rid of all the poor dead bodies, instead of investigating how that mess really happened, the authorities got rid of the evidence by sending all the remaining steel beams of those buildings to China by ships!

I could really imagine, with my mind's eye, how the Chinese people received the steel and probably analyzed the steel to find out what kind of explosive the "Americans" had used to make those buildings collapse.

That was too much for me to take, considering that growing up in the upper floor of a police station with my father being the chief and the rest of the subordinates that investigated everything. Even for a car accident, they would take detailed measurements of how and why it happened.

Here in America people didn't even want to know the truth about why and how those buildings collapsed!??? And all those thousand of lost lives? Why? It had to be a dark, very dark agenda. They silenced and got rid of anything that would give a clue. They left no evidence.

In 2006 I discovered that the government had lost track of three trillion dollars. An Investigation Audit was ordered to find out how all that money disappeared. The audit office was in one of the towers... All the people that worked in the audit office died in the collapse of the buildings.

Underneath the third building that collapsed, there were several trucks full of gold ingots. The trucks left that building, when everyone's eyes were looking at the first two towers collapsing. Those trucks drove off, and shortly after, the third building also collapsed.

The gold was there because the US government had to return it to China.

During World War II the Chinese government was afraid that the Japanese were going to invade China because of all that gold they owned, so the American government made a deal with the Chinese that they would keep their gold for them for 50 years and then return it.

When the 50 years ended, the Chinese asked for their gold back, but the Deep State/USA corrupt government didn't want to return the gold to China. So there was a lawsuit in closed doors and the US government lost. The trucks with the gold were in the basement of the third building, ready to be returned to China.

The gold was never found and most of the US population until today doesn't know that there was a third building that went down on that September 11.

———◦≺◆≻◦———

The word Illuminati started to appear more often on the internet, along with Cabal, Zionist, Rothschild, Rockefeller, the Deep State, and the Khazarian Mafia.

Then the word corporation conglomerate made me curious and I was wondering...

Who was in charge of these conglomerate corporations?

36

They had made an intricate web of interlocking different companies and names and made it very difficult for us to unravel the web of deceit of those who were implementing the One World Order.

In 2018 I found out that Monsanto-Agent Orange- was attached to another well known sinister company named Blackwater. Also known as Xe or Academy, this is a private Jesuit army obeying the dictates of black nobility. This army operates outside of the law and it is also controlled by people, both within the US government as well as outside, mainly from the Vatican State.

Blackwater is a religious army serving the Pope of Rome, through the Order of Malta. The Order of Malta is considered under international law to be a "sovereign entity" with special diplomatic powers and privileges. Like Blackwater the order of Malta is "untouchable" because it is at the heart of the elite aristocracy. This mercenary army "Blackwater" (with three names) is the largest mercenary army in the world and it was sold to the multinational Monsanto. More recently Monsanto was purchased by Bayer…Pharma has the army now?
This is the link where I got part of this info: worldtruth.tv

On September 9 - 2022 Q) The Storm Rider, Official page.

I read a paper with the title: Deep State World Collapse.

The recent announcement made by the Vatican of all the activities of the Central Banking System in the world in almost all countries must be returned to the Vatican Bank. This includes all paper money/coins and also gold, silver, precious metals and minerals. All documents and debt collections related to the World Bank, real estate markets, loans, marketed services and debts must be returned to the Vatican between 1 and 30 September 2022.

This important Vatican announcement was made shortly after the announcement of the death of the Queen of the United Kingdom. Those who know about the great awakening movement knew that the queen had been executed years before.

The announcement of her death was a warning from the generals of the Military Alliance to let everyone know that the end of the Deep State Cabal was approaching.

This was recorded and sent to compromised [DS] leaders and Black Generals and was a direct warning to the Vatican and the Khazarian mafia.

Now the announcement of the queen's death ensures the [Collapse] of the world Cabal Deep State regime. The 14 countries controlled by the Queen: Australia, New Zealand, Canada, etc. all those countries will stop paying the Royal regime of the Monarchy, installed by the Cabal, which means that the possession of the British Empire and of the 14 countries around the world will stop all funds related to global money laundering operations via the UK, Canada, Ukraine,

Russia, Germany, Austria, Switzerland, New Zealand, Australia, South Africa etc.

Now, the Vatican is fearful of losing its assets and the control it has around the world (they) have asked State Street, Vanguard and BlackRock to return the assets to the Vatican Bank as the world collapse continues. In the Vatican, the leaders of the Cardinals officially loyal to the Khazarian mafia, the leaders of the Jesuits and the Knights of Malta (imprisoned within the Vatican walls cannot leave the Estate) have been locked up.

Military operations silently took over the Vatican empire's chain of command. That chain of command of the military order of Malta, controlled and had installed the UN, under the control of the Kazarian mafia.

Now with money running out and rapidly disappearing across European countries ... NATO and the UN are in an internal battle to raise money and the internal struggle is out of control. The members of the "cabal Zionist" are removed within the Vatican City.

This is causing a leadership vacuum in the G7 group of countries as the entire socio-economic system collapses.

In early 2022 the Italian government was quietly infiltrated, or overthrown and replaced with a REGIME that was installed by some of Italy's wealthiest families that have banned together and united with the white Hats because they were shocked that the country had been sprayed with frozen spike proteins and the first BIO-WEAPONS released in their country which started the global pandemic ... Thousands of their friends, family and children died in the release of the organic weapon Covid.

This coalition of wealthy elite and White Hats families is compliant with Operation White Hats which ensured the downfall and arrest of the Sovereign Military Order of Malta in the Vatican.

The serpent symbolism is all over the Catholic Religion. In the Vatican, in St. Peter basilica the Pope sits in the mouth of a serpent, as the tongue that preaches deception.

<hr>

In 2023 I discovered:

While all the deception was starting to show his real colors this tunnel was uncovered.

THE ASTONISHING 1500-MILE TUNNEL: VATICAN TO JERUSALEM REVEALS MIND-BOGGLING GOLD STASH!

In a shocking revelation that would make even the wildest conspiracy theorists' jaw drop, an underground tunnel stretching an unbelievable 1500 miles (241 Kilometers) from Vatican City to Jerusalem has been unearthed.
But that's not all! This incredible find also boasts a mind-bending treasure trove of gold, so vast that it required an armada of 650 airplanes to transport it out!

Nested beneath the hallowed grounds of the Vatican, an inconspicuous tunnel snakes its way beneath oceans, mountains, and lands, connecting two of the most revered cities in the world: **Vatican City and Jerusalem**. What lies within this labyrinthine marvel is not short of astounding. The sheer scale of this can hardly be grasped.
Read the full article HERE:
https://amg-news.com/the-astonishing-1500-mile-tunnel-vatican-to-jerusalem-reveals-mind-boggling-gold-stash-video/

Now, you may be wondering how such a monumental feat remained hidden for so long. The answer is simple: a veil of secrecy, upheld by powerful entities invested in maintaining the status quo. For years, whispers of a clandestine network encompassing religious institutions, influential figures, and shadowy organizations have circulated in hushed tones. This astonishing discovery lends credence to those who have long suspected a web of secrecy woven through the very fabric of our world.

Now, November 5, 2022, Italy receives Russian gas and its imports from Russia have doubled as Spain has also started importing it from Russia.

The LIGHT shines in the darker parts of the Satanic Kabala's programs and regime.

Control over money printing within the United States remains in the hands of the Nazi faction and the Khazarian mafia faction.

No matter what is happening, know that the military Alliance is quietly driving events.

Not only here in America, but all over the world, corrupt governments are falling… collapsing.

It's like we're watching a horror movie where the scariest/difficult part of the movie is near the end when it seems all hope is lost...

In June 2022, Dr Mercola Newsletter
gave me some more answers about the web of deceit of corporation conglomerate:

Big Pharma and mainstream media are largely owned by two asset management firms: BlackRock and Vanguard.

Vanguard and BlackRock are the top two owners of Time Warner, Comcast, Disney and News Corp, four of the six media companies that control more than 90% of the US media landscape.

BlackRock and Vanguard form a secret monopoly that owns just about everything else you can think of too. In all, they have ownership in 1,600 American firms, which in 2015 had combined revenues of $9.1 trillion. When you add in the third-largest global owner, State Street, their combined ownership encompasses nearly 90% of all S&P 500 firms.

What does the New York Times and a majority of other legacy media have in common with Big Pharma? Answer: They're largely owned by BlackRock and the Vanguard Group, the largest asset management firms in the world. Moreover, it turns out these two companies form a secret monopoly that owns just about everything else you can think of too.

Stocks of the world's largest corporations are owned by the same institutional investors. They all know each other. This means that 'competing' brands, like Coke and Pepsi aren't really competitors at all since their stock is owned by exactly the same investment coGet ready to delve into the mesmerizing depths of this enigmatic and controversial discovery.
In a world where secrets hide in the shadows, this incredible story will leave you questioning everything you thought you knew. Companies, banks and in some cases governments.

The smaller investors are owned by larger investors. Those are owned by even bigger investors. The visible top of the pyramid shows only two companies whose names we have often seen... They are Vanguard and BlackRock.

The power of these two companies is beyond your imagination. Not only do they own a large part of the stocks of nearly all big companies but also the stock of the investors in those companies. This gives them a complete monopoly.

"A Bloomberg report states that both these companies in the year 2028, together will have investments in the amount of 20 trillion dollars. That means that they will own almost everything"

The assets of BlackRock alone are valued at $10 trillion. Making this circle of power even smaller, Vanguard is the largest shareholder of BlackRock. And who owns Vanguard? Due to its legal structure, ownership is difficult to discern. It's owned by its various funds, which in turn are owned by the shareholders. Aside from these shareholders, it has no outside investors and is not publicly traded. That said, many of the oldest, richest families in the world can be linked to Vanguard funds, including the Rothschilds, the Orsini family, the Bush family, the British Royal family, the du Pont family, and the Morgans, Vanderbilts and Rockefellers.

I predict that many years before the date 2028 these companies are not going to exist any more...The truth will remain in the history books.

Ps: Few years ago Google had removed Dr Mercola business from their search engines...
I wonder why Google people don't like his way of exposing the truth?

Well... evil does not want the truth to be exposed...

The book of Revelations has started to flash strongly in my mind since the year 2008 and I realized that the elite powers of the world were moving things to bring Humanity to their knees soon. Eventually I started hearing about concentration camps being built throughout the USA, They were called FEMA camps.

Around 2018, I started to see a lot of videos of military convoys moving through the country. These videos were posted on YouTube by everyday, concerned people. The corruption in the government was immense. I started to think that the time of tribulation was in full swing!

PS: Zionism are the pretending Jews, which is not Judaism.
There are 9 millions of decent Jews in America, who are not the problem. Zionism is the criminal element that runs Israel that has sabotaged the US Government.

President J. F. Kennedy warned us!

He was assassinated on the 22nd of November, 1963, ten days after he made this speech at **Columbia University on Nov 12, 1963.**

"The high office of the President has been used to form a plot to destroy the American's freedom and before I leave office, I must inform the citizens of this plight."
"For we are opposed around the world by a monolithic and ruthless conspiracy that relies on covert means for expanding its sphere of influence, on infiltration instead of invasion, on subversion instead of elections, on intimidation instead of free choice, on guerrilla by night instead of armies by day. It is a system that has conscripted vast human and material resources into the building of a tight knit, highly efficient machine that combines military, diplomatic, intelligence, economic, scientific and political operations."

President Kennedy knew the truth, the truth that took us many more years to uncover.

The President had intention to close the Federal Reserve as soon as would get back to Washington DC White House. Instead he went back in a coffin, but in the coffin President Kennedy's head was missing…The Kazarian Mafia, Deep State, One World Order etc.. kept it as a Trophy…

I was a teenage student at that time. I lived in Italy, I didn't understand much of the English language but I remember how I cried at that loss. He had been my hero even if he wasn't the president of my country.

I followed the Italian news on my TV for so many days after his death. I felt deep in my being that there were a lot of lies and a charade of nonsense created by those who were "supposed" to investigate and discover what really happened, that terrible day!
I could feel deep inside of me that the stories perpetrated in the news were not true. I am a sensitive intuitive.

Until today the truth is still hidden but there are talks now that many more shooters were there that day in Dallas who killed the beloved president, from different directions. Definitely the speeches that President Kennedy gave in the last weeks of his life were a real threat to the Bankers-gangsters of the Deep State, Kazarian Mafia, because President Kennedy had also said that he was going to close the Federal Reserve as soon as he was back in Washington DC that week.

And I realized that this was one of the real reasons why he was killed. (Money Talks and...) Now the Control over the printing of money within the United States remains in the hands of the Nazi faction, the Khazarian mafia faction and Rockfellers. In the year 2,000 and still now.

On January 22, 2023 From a video of Dr Mercola with Whitney Webb's Book "One Nation Under Blackmail" which is now two books.

https://takecontrol.substack.com/p/one-nation-under-blackmail

Whitney says: I found out that there is a lot of merging happening between Big Pharma and Silicon Valley. They are having big funding from Intel IQT (which is CIA Venture Capital). Their effort is to push this Technocratic Transhumanist System.

More people should pay more attention to Transhumanism because this is the new IGENIX of Health Care. Whitney also says that between 2023 and 2024 we are going to see more of the Central Bank Digital Currency (CBDC). They are going to have a very advanced program which will be framed first as voluntary and then once enough people start using it it will become the only form of legal tender in use. And this is very Bad.

We must stop it and do whatever we can not to use this system of currency.

On January 23, 2023 Benjamin Fulford Weekly Geo-Political News and Analysis.

We are also being told by Polish intelligence that the United States has set a cut-off date for US dollars legal tender restrictions that will expire on January 31, 2023. The restrictions mean that any US dollar bills printed before 2021 will no longer be accepted as legal tender anywhere in the world.

This is a desperate attempt by the Rockefellers to prevent bankruptcy by forcing everyone to use digital central bank currencies they plan to issue, MI6 sources say. Instead, the world is going to cut off the Rockefellers by refusing digital currencies they produce after January 31st, multiple sources agree. The bankruptcy of the Federal Reserve Board means the entire US financial system is imploding.

On April 22-2023 Dr. Mercola Newsletter

THE WAR WAGING AGAINST FINANCIAL FREEDOM

The financial guru Catherine Austin Fitts. Publisher of Solari Report- Who has warned that CBDC is part of a plan to end all currencies. If that happens a slavery system, steeped in ideologies of transhumanism and technocracy, will be ushered in.

"We're watching events that are within a framework, which is very engineered and planned "Fitts says "At the root of what's going on today is there is a group of people who are trying to totally centralize control of all financial transactions on the planet-100%- using digital technology."

Why are globalists promoting central bank digital currencies, or CBDCs, so heavily?

With cash we don't know, for example, who's using a $100 bill today. We don't know who is using a 1,000-peso bill today. A key difference with the CBDC is that Central Bank will have absolute control on the rules and regulations that will determine the use of that expression of Central Bank liability and also we have the technology to enforce that"

This is a very rare moment when a central Bank is telling the truth detailing how central banks can enforce rules centrally because it's no longer your money, it's -our money - and we can set the rules on how you can use "our" money", she says.

CBDCs will rapidly usher in an era of taxation without representation, leading to the end of liberty. By granting complete control of individuals' financial transactions to central bankers, CBDCs allow the government to maintain complete control.

Without financial transaction freedom, the ability to transact with whoever you want, with any purpose, there will be no freedom. Fitts says. This is a key reason why we need to preserve cash. Right now, there's a war going on between decentralizers who are trying to preserve financial transaction freedom, and centralizers who are fighting among themselves over who will be in control of the system.
"And, of course, every effort is being made by the corporate media to turn the rest of us who are trying to fight for transaction freedom to divide and conquer. And so it can be very confusing to watch this if you don't see the gist of the main game." Fitts explains.

CENTRAL BANKERS HIDING BEHIND HEALTH INFRASTRUCTURES

Part of what is going on behind the scenes is what Fitts refers to as the Going Direct Reset-"the wholesale plan" of the Great Reset, which has been packaged for "retail sale" to the masses. The Going Direct Reset is detailed by John Titus on the Solar Report, but it involves the massive amount of money- 3.5 trillions over a few weeks- injected into the economy in 2020.

The money was largely used in a way to build out only certain sectors- like space, the smart grid and health infrastructures, while starving others. This is another fact of gaining control and also involves the rollout of digital passports under the guide of keeping the population healthy and safe.
"Essentially, build out the infrastructures of control. Get everybody on an electric grid... and we see this dance between finance and health care. If the Central Bankers had to do all the centralizing control with money they would end up with everybody coming at them with pitchforks, and so they hide behind health.

We see this use of the health infrastructures basically to build the train tracks of control.
So the central banks, for example, to do CBDC they need a digital ID. EWell, you get that because you're trying to make everybody safe, right?... So we've had this dance during the going direct reset and using Health to justify more central control.

WAS THE SVB BANKING COLLAPSE DELIBERATE?

(I Rosanna the moment I was informed of the Silicon Valley bank (SVB) collapse, I knew that that was another nefarious move of the criminal Bankers... and days later here I read the details of the real truth, uncovered by Catherin Agustin Fitts and I learned why they collapsed the bank)

During the Pandemic, you had the Fed pumping a massive amount of money into the economy, while one-third of half of U.S. small businesses were shut down in the name of public Health. This wreaked havoc with people's loan portfolios, but what caused SILICON VALLEY BANK (SVB) the 16th largest in the U.S. collapse?

" At Silicon Valley Bank (SVB) " Fitts says, "You have the biotech and life sciences and the Tech IPO pipeline that literally sort of explode in a bubble, and then the bubble's over it kind shuts down."
That was an aspect. Meanwhile, Fitts says, "49% of the small businesses in San Francisco shut down" So if you're SVB loaning to just small businesses in the Silicon Valley area...that could be as much as half your loan portfolio." However, ultimately Fitts believed the collapse was a deliberate takedown-not the result of a traditional bank run.

"We had a takedown at SVB. There's a game going on and...what it turned into was an effort by a variety of players to panic everywhere into believing... that this was going to turn into a wider contagion run. Now if you look at the numbers on the banking System, if interest rates continue to stay high for a long time and a lot of banks run a negative arbitrage, that's a problem...To get complete control you have to kill small guys. You've got to kill the small farms. You've got to kill small businesses and you've got to kill the small banks.

So we use the health care game to shut down the small businesses and the farms-not because they're less productive but because the guys running the game could pick up huge market share and make a fortune stealing their businesses and picking their asses cheap…
You pump money into your pals and then shut down your enemies, and then your pals can go pick your enemies up for cheap…so we shut down the small farms and small businesses…and now we're ready to shut down the small Banks.

Now, if you're the top guys and you want to play this game, who better to shut down the small banks than panic on all the small bank's depositors and scare them and getting them to walk their deposits across the street to the criminals?…What they're trying to do is get all their neighbors' cattle to stampede into their carrals."

If the globalists take over, which is all but guaranteed if they control the financial system with CBDCs, they can institute worldwide slavery. But unlike in the past, technology now exists to keep track of people's every move and control their ability to live in the modern world if they don't obey.

Fitts says:
" The greatest most profitable business in the history of the world is slavery, more than mining, more than anything else. Slavery. That's what this is about. Digital technology solves all the bad problems they had with slavery last time. All those problems you can now solve with digital technology.

You can perfect your collateral- if I can implant a chip in You… then I can collateralize you. I can perfect my collateral as a banker and now we're off to the races. I can build a slavery system and with robotics, AI and automation, I can manage it."

HOW TO FIND A GOOD BANK

Fitts stresses that leaving the banking system isn't the answer - finding a good local bank, and stashing the cash in a variety of places, is. While you're there, let bankers know about the dangers of CBDCs. The Solari Report even has a template letter you can use to inform your bankers about the downsides of CBDCs. it reads, in part:

"it strikes me that creating a different, yet centrally fiat currency that can be created from thin air and manipulated by unelected central bankers does not promote U.S. financial stability or provide citizen with consumer and investors protections- except in the sense that totalitarian governments can be financially stable through the power of taxation without representation and the ability to micromanage and regulate the spending of families and small enterprise."

The bank or credit union you choose should not have a record of criminal behavior. Next, just as it's a good idea to get to know the small farmers growing your food, it's a good idea to get to know the people running your local bank.

The "No. 1 criteria is are they well managed? Are they well governed? Who owns, who manages, who's on the Board?" fitts says. "What is the quality of the people and their experience and expertise?" The second criteria to look for is a steady deposit base.

"You want to look at the deposit base, you want to look at the loan base, you want to look at the investment portfolio" according to Fitts. "And then you want to make sure that in the investment portfolio or in their accounts, they have enough short-term investments and cash…It's like a train. People get on and off, and you want them to have enough liquidity so they can handle the ups and downs."

Fitts encourages everyone to get to know their community bankers and credit union executives. "If you're concerned about your bank, go talk to your bank. If they're doing a good job, they will answer your questions with confidence".

HOW TO STOP CBDC

One of the top ways to stop CBDCs, in addition to ditching large. multinational banks in favor of trustworthy local banks or credit unions, is to use cash as much as possible, and not frequent shops that don't accept it.
You want to have your cash spread out, starting with keeping cash on hand in your home, ideally in a fireproof safe. Then you can expand to a safe deposit box at the bank, or investing in silver and gold coins. Meanwhile, talk to the people in your local community about why you're using cash.

"Start using cash, and as you use cash, talk to the small business that you frequent- small restaurant, small farms- and talk to them about how we can work together as customer and business to improve our endurance and resiliency and well-being," Fitts says.

Helping to build out your local food sources is also part of the solution, since having access to healthy food is critical to maintaining health and autonomy.

"Anything you build out your local food markets so that we can't get cornered and be dependent on our enemies for food, it's going to make a big difference to freedom," she explains.

There's still time to defeat the globalists and maintain life as we know it. So, rather than feeling defeated, recognize that the opportunity exists to win this battle, one action and one individual at a time…

"Understanding this can be completely overwhelming and depressing, but the thing I want to encourage everybody to understand is facing it… is the doorway that you walk through. And the grief you experience to get to the other side and realize there are solutions, but if you face reality, there are real solutions.

Control over money printing in the United States remains in the hands of the Nazi faction and the Khazarian mafia faction.

Whatever is happening, know that the Military Alliance is silently guiding events.

Not just here in America, corrupt governments around the world are collapsing.

It's like we're watching a horror movie, the scariest and most difficult part of the movie comes near the end when it seems like all hope is lost.

There are dark forces in every government in the world now.

There are more pandemics planned.

There are planned famines due to food shortages that are easily traced back to governments.

The Cabal's intention is to remove your power and dignity.

Be prepared and protect your home and life.

Know that things will get worse in many ways before they get better.

We must never give up! We have help from above!

We must realize that the energy we have within ourselves keeps us connected to the ETERNAL SOURCE OF ALL THINGS.

The power of LIGHT SOURCE OF GOD, these days we are all being flooded by an immense quantity of quantum WAVES OF LIGHT that come from the Central Sun of the Universe. The dark forces of evil cannot absorb this immense amount of LIGHT, they live in darkness and their power is weakening.

We need to focus our attention on maintaining our bodily form in a state of purification; that way we maximize our chances of completing the transmutation process that will give us access to higher levels of density and consciousness.

CHAPTER 5

FEDERAL RESERVE, BANKERS AND BANKS

The Federal Reserve is an engine of the Deep State; it has global connections, is international and is an economic shadow government. It runs the USA government. There are no "Reserves" in the Federal Reserve, it is a fake name.

The Federal Reserve was started by a secret society, the elite Bankers (now I know as a part of the Khazarian Mafia) primarily the house of Morgan, connected to the house of Rothschild with deep connections, society and dirty horrible evil practices.

The Deep State globalists want a Global economy, a one World Order and a one World Government, with a Central World Bank.

This evil idea started and has been going on since, before 1913.

Those who control the economy and our freedom are the world leading Banking Cartels, Morgan, Rockefeller, Rothschild, Warburg etc.

The chairman of the Federal Reserve is appointed by the President of the USA, the "Fed" appears to be like a real Federal Agency.

However there is nothing Federal about it. The Chairman is only a figurehead; the real hidden Authority Director is in New York and directs the show in secret.

The Federal reserve is not a government agency, it is a private bank internationally connected. There are 12 branch directors of powerful international banks and their money is laundered into the US economy. They control the US economy. Their secret deliberations and secret budget cannot be audited, so it cannot be determined where the money goes.

Congress has no control or access.

The Stock market is a Scam.

Stock market is a scam, a tool of the Deep State - Secret Space Program used to relieve gullible beings of their hard earned fake Federal Reserve notes. It's all just a paper chase numbers in a system they control. Debit Slavery, Central Banks are not asset backed.

We, the living Being have been the Asset, the Collateral. Our Birth Certificate is a Slavery Bond Promissory Note.

The Stock Market crashed sometime. We are now on a Quantum Gold Asset Backed System. Your News Media has been pulling the wool over your eyes. As this is the Matrix Quantum Holographic Virtual Reality Simulation.

The Banksters have power to manufacture currency, print money like in the monopoly game, determine the value of the dollar, manipulate interest rates, and decide the economic conditions of America. They eliminate the freedom of US citizens with financial control of our purchases, our employment, our houses, cars etc. They own us like their slave possessions. The increase of totalitarian control is here, and it is attempting to take over the globe and Humanity.

"He who learns but doesn't think, is Lost! He who thinks but doesn't learn is in great Danger."

~~Confucius~~
722-479 BCE

All the Banks are owned by private cartel, organized crime groups that support kings and dictators. The people that are awake and wise know that the name of this cartel is the Khazarian Mafia, which has many different names…

DARPA controls INTERNET, BIG PHARMA AND BANKS.

This is the top of the pyramid, a crime group on our planet. They are deeply dirty minded against humanity. They own your religions, politicians, mainstream media and monetary systems.

Why are so many people not asking themselves, who are the ones that profit while sitting on the top of this pyramid of war, disease and human trafficking?

Too many humans don't understand that those on top of the pyramid also own your mind and the last thing they want is for you to take the red pill. All this information that you read is causing the Human Awakening is the "red pill".

For thousands of years these off-world controllers have done an incredible job of dividing humanity including financing both sides of all wars. Even our history has been one big manipulation.

Israel will expose the root of all corruption and evil in our religion system.

The revelation of the hijacking of scriptures by Satan many centuries ago. The Light of our planet was dimmed by the false light of darkness. The snake entered the Garden of Eden and went undercover into the snakepit.

The head of the snake is in the Vatican, the body of the snake formed the silk road leading to Wuan where it had spread its venom. Hiding into the snake pit of Israel where Satanism hijacked part of the scriptures and became the false light that fueled humanity in evil frequencies of fear, shame and guilt.

Separated from God and the Love we are, separated in Religion, race, politics, creating dark for centuries. We were separated by a small group who benefitted from the controlling of energies and the people.

All the Royal families in Europe and the Baltic, and Russian countries claim their ancestors of VOTAN or ODIN. Votan and Odin were a PENTAGON- SERPENT- SNAKE -REPTILIAN. 13 ROYAL families are REPTILIAN HYBRIDS who are shapeshifters posing as Humans.

The DRUSE bloodline of Jesus are descendants of JETHRO, the priest of MIDIAN in the Bible and the "Torah" (Exodus 2:18). The 16 presidents of the United States of America "Abreham Lincoln" descend from the Kahlooni family.

In 1855 ISIS was formed by the kings of Morocco and Libya (Hassan Family), together with the rotor British family, signed the Mohaddi law which was to kill the bloodline of Jesus Christ (Druse). 6 years later they merged with Skull and Bones: Rothschild Schiff, Rockefeller, sheriff, aka Bush, Kissinger etc.

1855- ISIS Formed by sanussi Family linked to the UK Royals (Kazarians) 1861- Merged with 322 skull n Bones (Kazarians)

1870-1930 BIG PHARMA (Kazarians)

1871-Act of England (secret constitution placed by secret societies) (Kazarians) 1912- Titanic / Olympic Sinking (who was on board? what really happened?) 1913- Federal Reserve

1945- 1959 Operation Paperclip (Mockingbird)

1948 ISRAEL formed (Kazarian/ Bolshevik government) 1949- MOSSAD= CIA formed.

Whistleblower General Flyinn exposed the Gulen terrorist network: Flynn discovered that the Obama administration was creating funding and arming Jihad actions taking place in Gaza.

Administration and by default of HWG Bush, Clinton * Obama with the help of CIA, Nato and his Jiadist cronies, were seeking the Top ASSAD in Syria.

* These treasonous administrations created funding for ISIS a. o.

* The US /Jiadist morphed into the Arab spring, destabilizing the Middle East and creating the orchestrated migration crises.

* Hillary Clinton took part in the launching of the Abram Spring, as secretary of state under the Obama administration.

* Flinn blew the whistle on the involvement in the Gulen Terrorist Network of Obama, the state department, FBI, CIA etc.

Don't be fooled/ those buildings fully collapsed in Gaza strip being hit from one missile is used in controlled demolitions.

Deep State operators inside both Israel and Palestine Camps have been infiltrated long ago by Mossad /CIA.

Israeli handlers are in fact Mossad, >> UK, M16, Rothschild >> CIA>

Dominion Servers have a lot to hide: from human trafficking, to Epstein creations, to Vatican banks. The Mossad/ Kazarian mafia controlling Ukraine connections from the Snake Pit of corruption by ISRAELI Elites run deep into United States MSM [CIA] control.

December 5--2023 From Benjamin Fulford.

*NOTE * Jesus was a real person with important super intelligence and his understanding of the true Spiritual energy that can bend reality and time.. Which means defeat death. TIME TRAVEL<   /

_unfortunately a lot of his true teachings were hidden and suppressed<
Now there are over 20,000 different denominations of Christianity and they are all fighting about who is right and wrong.

>The Roman/Christian wars created the first banking Systems and military intelligence and massive human trafficking networks. They stole most of Europe and large parts of middle east gold, silver, arts resources and historical documents and hid them under the Vatican vaults and caves...

To this day the Vatican museum and open vaults house over a trillion in arts and sculptures and jewelry and precious metals stones the public can see.. But what you can't see is the hundreds of trillions of artifacts hidden in the Vatican underground caves, bunker's and vaults that stretch over 50 miles underground.

The Satanic power that infiltrated the Vatican and pedophile world networks is connected to KHAZARIAN powers, Mossad, Cia, Rockerfellers and Rothschilds who helped MOSSAD, Cia, financed Robert Maxwell the father of Ghislaine Maxwell> and Jeffrey [EPSTEIN]

DRAIN THE SWAMP means draining the world of the SATANIC CABAL .

You are inside a powerful GREAT AWAKENING.

NESARA's lost dollars!

On January 21, 2024 I discovered that NESARA was to be announced on September 11, 2001 at 10:00 am by Federal Secretary Alan Greenspan.

However, an evil plan was orchestrated by former CIA Director George Bush Senior.

His family was handling $412 trillion which was the money humanity was supposed to access with NESARA. All those dollars had been entrusted to him and his family, because then they would be available.

This quote was written by George Bush Sr.: "Sera, if the American people found out what we did, they would chase us down the street and kill us."

signature: George M. W. Bush Sr

⊂·≺⊀◊⊁≻·⊃

NESARA will be announced soon, the money has already been moved, the Treasury Department already has it. The United States no longer uses fiat dollars, that money is dead, no longer has any currency.

The Federal Reserve was wiped out; all three-letter agencies, domestic and globally, are under the control of the White Hat military. There is nothing the deep state can control anymore. Evil is removed more and more every day.

⊂·≺⊀◊⊁≻·⊃

October 26- 2023 Benjamin Fulford article:

NESARA / GESARA: The Dawn of a New World

"The End of Poverty, The End of Debt, The Beginning of a New Golden Age!"

Prepare for a world where poverty, hunger, and debt are relics of the past, replaced by global prosperity and lasting peace for all!

NESARA Joins Forces with GESARA

From Benjamin Fulford

In a monumental shift, NESARA, a comprehensive economic reform plan for the United States, has united its destiny with GESARA, its global counterpart. This momentous announcement heralds a profound transformation that will impact not only the United States but also a coalition of 206 sovereign nations worldwide. The linchpin of this transformation is the new financial system enshrined within GESARA.

AGlobal Gold-Standard Monetary System

Once GESARA takes center stage, the International Monetary Fund (IMF) will declare the inception of a "global gold-standard monetary system." In this new era, all remaining fiat currencies will be exchanged for gold-backed currency, a significant step away from paper money. The march toward digital currencies will gain momentum under this revitalized financial system.

A Transition Rooted in Simplicity

To ensure a seamless transition, meticulous preparations have been made. The new financial system has been operational for months, securely hosted on a quantum server impervious to hacking or unauthorized access. Crucially, wealth proliferation is a cornerstone of this transformation. Newly minted wealth holders are more inclined to contribute to humanitarian efforts, ultimately fostering wealth for all.

Wealth-Building Mechanisms

This transformation might witness a shortage of skilled workers, which, paradoxically, is a wealth-building catalyst. The resulting surge in wages and salaries counterbalances the price drop stemming from tax reductions, sometimes up to 80% of final product costs, thus ushering in deflation. Lower energy costs, thanks to free-energy technologies, further contribute to this financial renaissance.

The Future: A Glorious Reality

Our future is poised to be a grand tapestry of innovation and abundance. Technologies once suppressed by the cabal, some dating back centuries, are finally being unleashed. For instance, the pristine waters of Antarctica will revitalize arid regions and infuse life into all flora and fauna. A world with personalized credit cards, Replicators that produce everything, and newfound awareness of the power of the mind to manifest our desires awaits. Real healthcare capable of rejuvenating our bodies, regrowing limbs or organs, and even reversing the aging process by decades is on the horizon.

Liberation from Financial Chains

Money and traditional banking, tools historically employed by the cabal to manipulate and control us within a debt-based economy, are slated for obsolescence. Coinage, however, will endure. With the elimination of national debts globally, taxes will be lowered for individuals and corporations. Instead, a flat sales tax of approximately 15% on new items will be implemented.

The Dissolution of the Global Elite's Pyramid

The current pyramid structure dominated by the global elite, governments, and corporations has reached its limit. This structure perpetuates class divisions and scarcity. True spiritual evolution arises when one's identity transcends material attachments. With the advent of free energy, advanced transportation, and Replicators accessible to all, true equality emerges. No one will be enslaved, people will pursue their passions with ample free time for reflection and creativity. Hoarding becomes unnecessary.

A Vision of Advanced Civilization

In advanced civilizations, concerns about food, shelter, and transportation are relics of the past. As NESARA and GESARA unfurl, we stand on the brink of a world where prosperity, freedom, and innovation thrive, and the human spirit soars unburdened by material concerns.

CHAPTER 6

SECRET SERVICES AND THE CIA

"I will splinter the CIA into a thousand pieces and scatter it into the wind"

~~ President Kennedy ~~

President John F. Kennedy gave a speech called Tyranny of Secrecy on **April 27, 1961.**

Here is a little piece of it.

"The very word "Secrecy" is repugnant in a free and open society...

And there is a very grave danger that an announced need for increased security will be seized upon by those anxious to expand its meaning to the very limit of official censorship and concealment."

Was President Kennedy talking about the Secret Services???

This is what I know about the CIA:

Operation Gladio

This operation is about when the CIA Secret Service took former German SS agents, changed their identities and sent them to Italy, Sweden and France to set up a false terrorist organization against what they perceived as Socialists and Communist governments. The CIA eventually staged false terrorist attacks and killed 490 people to make it appear as if it was the Italian government who made the attack in the month of August.

I was in Italy with my children, during the month of August there was a very big explosion at the train station in Bologna 41 years ago.
23 kilos of explosive were used, 83 dead and 200 hospitalized. It was at the time of the year when all the Italians that live in the north of Italy traveled south for their vacations, many go to the sea, others to the mountains and others go visit relatives that live in the south.

There were other explosions, in other places not at the same time and month, like the explosion in Piazza Fontana where 17 died and 88 went to the hospital. I was in America at that time and I don't remember it all, I know that was an extended, very bad, period of time that caused the death of all those 490 people.

The CIA also had connections with the Sicilian mafia for years, and started pushing drugs into the USA in the inner cities where there is a lot of drug abuse.

The CIA made millions of dollars and used the Vatican's secret bank to launder the money and install false terrorist attacks in various places.

In the 1970s/1980s they went to South American countries and got rid of various leaders of those countries and placed puppets in their place. Then they smuggled drugs from there into the USA.

When I think about Afghanistan and all the opium that was growing over there and how many atrocities happened in that country… It made me wonder if the opium was another avenue to make millions of dollars from its sale by the Luciferian Cabal/CIA.

Lately, I started thinking that probably that same source of opium has been used by the Big Pharmaceutical companies to make painkillers that the doctors have been prescribing all over the USA and the rest of the world to make their patients "pain free" and addicted. So many young and old people eventually died from addiction to those "painkillers."

The CIA headquarter is located in Switzerland, not in the USA.
All roads don't lead to Rome, they lead to Switzerland. Swiss guards are the guard of the Vatican. The Bank of International Settlements is in Switzerland, "Rome" is in Switzerland.

All the CIA does is not only illegal here in the USA, but they continue to do it all over the world. They break the law with impunity all over the World.

Later I learned that the CIA controls Hollywood and Mockingbird media staged agendas and connections for child trafficking. It has been creating software that is used by the CIA, police, and elites to hunt down children through face recognition.

All the way back in 1961 President Kennedy was talking about "Secrecy"
Official censorship and concealment!

———◦⊰◆⊱◦———

This morning March 18, 2022, I read on telegram:

Q) The storm Rider /official page

News from the Australian Special Forces Commander Riccardo Bosi stationed in Ukraine, who said: "Ukraine is the head of the snake (the Deep State) and Vladimir Putin is taking the Head off."

"Ukraine has been the center of the globalists for decades… the CIA has been working in Ukraine for 70 years" The commander continued "Ignore all the chatter about nuclear war and Russia's attempts to take over the Globe. It is completely the opposite!"

I, Rosanna have to say this now: We truly are witnessing the systematic destruction of the old guard. "The head of the snake is being chopped off."

The fake news said the CIA has admitted to having military personnel in Ukraine and are training groups of soldiers and Ukraine Nazis for a war against Russia, and for creating proxy wars with the east Ukraine sector for the past 8 years.

In reality here below is the truth:

As Russia was planning to release the documents and footage of the CIA training operation through Ukraine, the Pentagon panicked and released their story first…portraying the CIA as heroes training Ukraine military forces.

The truth is that the CIA has been training Ukrainian West Nazi Forces since the end of World War II. Between 1945/1950 and covered it up.

These Nazis (taken out from Germany at the end of World War II) never had to face Nuremberg trials for war crimes against humanity, because it was deliberately silenced from the Nuremberg trials.
The whole thing about west Ukraine creating a Nazi regime was financed and backed by Hitler, Prescott Bush (which was the father of President Bush senior), Rockefeller and the 3rd Reich regime.

⟶•≺ɐ◊ɒ≻•⟵

And now the mainstream media "fake story" is painting the CIA as heroes???…

Why does the western media want you to hate Vladimir Putin and Russia?

Where are the responsible journalists to report the real truth?

I have the feeling that the good reporters are systematically silenced or killed.

⟶•≺ɐ◊ɒ≻•⟵

On April 19, 2022 I continued to read Q) The Storm Rider /official page:

How do you expose the Deep State inside the Pentagon…CIA/UN/NATO/Dark (DS) Generals? US and World private Military/Rogue/Contractors?

PUTIN knew tens of thousands of US military were in UKRAINE illegally training and arming the MILITARY & militias.

TRUMP, XI, MODI, BIN Salman, PUTIN they all knew (DS) Military in US was arming Ukraine, Kazakhstan and training for the ultimate aim to attack Russia. (This has always been the goal of the 3rd and the 4th Reich. WEF, WORLD BANKS, CABAL…etc,

THE DEEP STATE GOAL of several assassination attempts on PUTIN-TRUMP-XI-MODI-BIN SALMAN -all failed. So how do you expose the DEEP STATE, PENTAGON/CIA/UN/NATO?

You bring them ALL into the brink of World War…and then you COLLAPSE the FIAT System… (THE RATS in the PENTAGON, CIA, NATO, ETC. ETC…jump ship, lose their protection when the money's gone- collapsing-) White hats, double agents/moles/ infiltrators and the [DS] ops began destroying their operations from the inside out… From the outside you destroy the world Ukraine money laundering scheme…

65

MAJOR PANIC IN THE HOUSES OF PENTAGON, CIA, FBI, 3 letter agencies.

May 4/2022   Q) The storm Rider /official page

Hundreds upon hundreds of millions connected to 80 years of trillions of $$$

Deep State funneling money into UKRAINE through CIA BLACKROCK money laundering system is all connected to…

Gates Microsoft, Facebook, YouTube, Google, and BlackRock/Vanguard

The Military are sending messages loud and clear.

It's only the beginning.

Billions aid package to Ukraine

350 millions in February
800 Millions in March
1.3 billions in April,

For the first time in history…a world AWAKENING on MASS scale is seeing the connections and sharing the information worldwide and growing.

It's becoming clear… it was the democrat's plan all along to push the U.S. into war and cover up Ukraine money laundering/bioweapon labs/human trafficking Networks…

Was this the Deep State's big plan… because they knew White Hats Military were getting ready to drop the HAMMER?

Everything happening has to happen… (Came to the very edge of darkness)
The hundreds of billions Biden is sending to Ukraine.
HRC, Obama's shadow government, Deep State sinister ops to create a World War…

"EVERYTHING IS BEING DOCUMENTED"

BlackRock CEO
Letter to shareholders

"As I write this letter to you, the world is undergoing a transformation: Russia's brutal attack on Ukraine has opened the world order that had been in place since the end of the Cold War, more than 30 years ago. The attack on a sovereign nation is something we have not seen in Europe in nearly 80 years, and most of us never imagined that in our lifetime we would see a war like this waged by a nuclear Superpower. But the Russian invasion of Ukraine has put an end to the globalization we have experienced over the last three decades.

It's impossible to predict precisely what path this war will take. BlackRock is focused on monitoring the direct and indirect impacts of the crisis and working with our clients to understand how to navigate this new investment environment. Implication of Russian invasion of Ukraine, for companies, countries, and our clients."

(This letter is a way of letting their shareholders know that if something happens… its Russia and the Pandemic). The billions that Biden is stealing from the American people (Taxes) and funneling to Ukraine are actually going to BlackRock.

Word in the CABAL'S: BLACKROCK has been highly infiltrated by the White Hats… and the FORCED PANIC scenario is running through the channels of the shareholders and Collapsing Banks. They can't keep up the charade of the Fake MSM MONEY MARKETS/FALSE STOCK REPORTS any longer. PANIC INSIDE BLACKROCK.

⊂·≺◈◈≻·⊃

Many years ago before the pandemic began, I found out that there is an intense network of underground tunnels and cities all over the world, kept secret. The pedophiles and the Secret Service use them for moving the human trade and more... The tunnel system underground, throughout the world, is now being exposed. In the US there are about 129/130 known underground facilities and many more all over the world. The people involved in building and running these underground cities are operating outside of the law.

In New York City there are underground tunnels hidden from the public that connect NYC to Philadelphia to DC. All this was connected to the human trafficking system of Epstein and other elites.

There are a lot of indictments, hidden courts, where trials are happening now, and military tribunals are taking place. The collapse of the Deep State is in free fall. But the media doesn't talk about it; they are too busy spreading lies, fear and propaganda.

We are now thinking that everything is moving too slowly and it is impossible to be able to get rid of the immense amount of dirty evil on Earth, but half or more of the population on Earth is already awake and a lot is being done in silence by the resistance, the White Hats and the Light forces.

Yes, there is a lot to do.

It is like attempting to turn the course of an immense ship, at first it is slow moving, but when it is pointed out to the right direction it picks up pace. So we are now picking up pace, even though the news media doesn't talk about it. They are going on with their propaganda to continue to instill fear in the ones not awakened.

"The wheels of Justice turn slow, but when they turn they are powerful."

~~Unknown~~

There are a total of 13 families that appointed themselves as the owners of the Planet and humanity. Each family has his own group of ownership.

These 13 Royal families are reptilian hybrids who are shapeshifters, posing as humans.

Five Cabal families control all the mass media in the entire world. Every nation gets the same made up BS. News, full of propaganda and false fear information promoted by the mainstream media. They are lies which are only fabricated to promote fear and chaos, and extort money from the population for "humanitarian purposes." This agenda is expedited especially now that their bank gold is reduced and the fiat money is dwindling down because the dollar is losing power.

The corporate giants are only interested in profit, on fostering their brand and gathering information on all that we do, so they can keep on manipulating us. This is a corporate controlled dictatorship system, similar to Fascism.

Those who control the money, control the world. But the World is changing and Humanity is waking up from the dream and eventually the corporate propaganda will not work any more in our deaf ears.

Many Dark military bases underground and under the sea have been cleared by the resistance movement which is aligned with the Light Forces. But the mainstream media full of fake news did not talk about this.

It is hard for us to admit to ourselves the real state of affairs, with all the lies and propaganda that they are throwing at us.

⸻◦⟨◇⟩◦⸻

-When you sell weapons you control armies.

-When you control food you control Society.

-When you control seeds you control life on Earth.

-When money is your master then your conscience is no longer your master.

~~Vandava Shiva~~

⸻◦⟨◇⟩◦⸻

July 7, 2022 by Dr Mercola:

- The "anti-disinformation" industry has nothing to do with protecting a gullible public from information that might cause them to make bad or unhealthy choices. It's about creating and directing a narrative for the purpose of controlling the population and hiding truths that might overthrow the ruling cabal and its plans for a one world government.

- In 1948, the CIA's Office of Special Projects launched Operation Mockingbird, a clandestine CIA media infiltration campaign that involved bribing hundreds of journalists to publish fake stories at the CIA's request.

- During the Cold War, CIA propaganda disparaged communist ideologies. Today, it promotes radical ideas that bring us closer to The Great Reset — which is based on a technocratic economic system — instead.
  Most of the organizations claiming to promote truth and counter disinformation are in fact doing the exact opposite. The latest and most blatant example of this was the Biden administration's "Ministry of Truth" — the Disinformation.

- Governance Board, set up by the U.S. Department of Homeland Security (DHS)

- Evidence shows scholars and academics who speak out against the establishment narrative on the conflict between Russia and Ukraine are being targeted by media personalities working hand-in-hand with the intelligence apparatus.

"In 1976, Senator Frank Church's investigation into the CIA exposed their corruption of the media ... The tactic was straightforward. False news reports or propaganda would be provided by CIA writers to knowing and unknowing reporters who would simply repeat the falsehoods over and over again."

THE MEDIA IS MORE CONTROLLED THAN EVER

It's the Opposite of What They Claim it is, it's no small irony that most of the organizations claiming to promote truth and counter disinformation are in fact doing the exact opposite. The latest and most blatant example of this was the Biden administration's "Ministry of Truth" — the Disinformation Governance Board, set up by the U.S. Department of Homeland Security (DHS).

Calling someone a 'conspiracy theorist' is a strategy aimed at silencing dissent in general and truth in particular, plain and simple.
https://www.youtube.com/watch?v=VKZ5ksMg2Q4&t=720s

—◦◦◦<◦◦◦◦—

January 27, 2023 Dr. Mercola About INTERNATIONAL NEWS COVERAGE

Most of the international news coverage in Western media is provided by three global news agencies — The Associated Press (AP), Reuters and Agence France-Presse (AFP)

The technocratic apparatus uses these news agencies to proliferate certain narratives while burying or 'debunking' others.

Until or unless at least one of these news agencies sends out a notice, national and local media are unlikely to report on an event. Even photos and videos are typically sourced directly from these global news agencies. This way, people hear, see and read the exact same message everywhere.

Intelligence agencies and defense ministries are well aware of the power of these news agencies and use them with regularity. In 2009, then-president of the AP, Tom Curley, let it slip that the U.S. The Pentagon has more than 27,000 PR specialists that spin up stories, and an annual propaganda budget of nearly $5 billion.

The rest of the technocratic apparatus uses these news agencies in the same way and for the same reasons — to proliferate certain narratives while burying or "debunking" others.

The Department of Homeland Security's Cybersecurity and Infrastructure Security Agency (CISA) is partnered with a censorship consortium called the Election Integrity Partnership (EIP). Through this consortium, the DHS is illegally censoring Americans.

⊂●⫶⟨⊹⊹⟩⫶●⊃

Today July 10, 2022.
THE SHOCKING TRUTH ABOUT THE GLOBAL DEPOPULATION AGENDA
On Dr Mercola
   I watched the ½ hour movie "Infertility: A Diabolic Agenda" It is free: infertilitymovie.org

- "Infertility: A Diabolical Agenda," produced by Dr. Andrew Wakefield and Children's Health Defense, details the World Health Organization's intentions to produce an anti-fertility vaccine in response to perceived overpopulation, and how such vaccines have been used — without people's knowledge or consent — since the mid-'90s

- The WHO has been caught more than once deliberately deceiving women into thinking they were vaccinated against tetanus, when in fact they were being sterilized.

- The film clearly illustrates the depopulation agenda is not a conspiracy theory. It's reality, and it's happening worldwide. The HPV vaccine and the COVID shots also have adverse impacts on fertility that are being ignored

- In the decade after the rollout of the HPV vaccine, the teen pregnancy rate dropped by 50%

- While VAERS is the only publicly available system to assess COVID jab injuries, the U.S. government has at least 10 other reporting systems they're not sharing data from. Children's Health Defense is filing Freedom of Information Act (FOIA) requests for the other systems to get a better idea of the scale of harms, but VAERS and anecdotal reports alone suggest the scale of injuries and deaths is enormous. Data from insurance companies around the world also confirm that "Infertility: A Diabolical Agenda" is Wakefield's fourth film. The first was "Who Killed Alex Spourdalakis?" followed by "Vaxxed" and "1986: The Act."

This latest film details the World Health Organization's intentions to produce an anti-fertility vaccine in response to perceived overpopulation, and how such vaccines have been used without people's knowledge or consent since the mid-'90s.

⟞⟞⟨⟨◆⟩⟩⟞⟞

July 16, 2022 - Will 100 million people die from Covid Wax by 2028?

David Martin, Ph.D. presents evidence that COVID-19 injections are not vaccines, but bioweapons that are being used as a form of genocide across the global population.

The spike protein produced by injections of COVID-19 is a known biological agent of concern. Martin believes the number of people who could die may have been revealed in 2011 when the World Health Organization announced their 'decade of vaccination'
The goal for the vaccination decade was a 15% population reduction globally, which would be about 700 million deaths; in the US, this can amount to between 75 million and 100 million people dying from COVID-19 injections.

When asked what time frame these people might die in, Martin suggested that "there are a lot of economic reasons why people hope it's between now and 2028."

Projected illiquidity of Social Security, Medicare, and Medicaid programs by 2028 suggests that "the fewer people in these programs, the better"; Martin believes this may be why people aged 65 and older were first targeted with COVID-19 injections.

In March 2022, Martin filed a federal lawsuit against President Biden, the Human Services Department of Health and the Center for Medicare and Medicaid Services alleging that injections of Covid-19 turn the body into a biological weapons factory, which produces spike proteins. Not only is the term "vaccination" misleading when referring to Covid-19 injections, it is inaccurate since they are actually a form of gene therapy.

"And not only will we not be sued for defamation or misinformation, we are actually holding people accountable for their domestic terrorism, their crime against humanity, and their history of weaponizing the coronavirus that goes back to 1998." says Martin.

⟞⟞⟨⟨◆⟩⟩⟞⟞

The government doesn't want us to see their fingerprints on anything. Everything is going out through the trust of the party that we are supposed to believe in, because no one trusts the CIA anymore.

They are causing record gas prices, record food prices and attempting to doome an entire generation to hopelessness. This type of economy is used as background noise to keep us in line, to make us consent. Most humans don't have the time and energy to investigate what is really the truth and who is beyond all this smoke and evil…because they have to make a living and pay the bills and taxes, they are in survival mode!

━━◗•⟨╂◆╂⟩•◖━━

November 21-2022 Benjamin Fulford NEWS

Since you see the World Economic Forum in the list, it turns out that Klaus Schwab Rothschild is just an old servant of Rockefeller,
These people are the Babylonian tyranny known as the "rules-based world order". Rockefeller Jr. was born to rule this nightmare created by John D. Rockefeller the first, who is pictured below.

JOHN D. ROCKEFELLER

My name is John D. Rockefeller. I started up the pharmaceutical industry. I worked to ban hemp. I hijacked the education system, co-founded the Federal Reserve and the United Nations, and helped found the Council on Foreign Relations. I was thought to be a hero entrepreneur, but in reality I was one of the greatest enemies of the American people.

Rosanna adds: also an enemy of all humanity in the rest of the world.

Many more people are responsible for spreading carcinogens and sterilizers in so many everyday products. That is why worldwide sperm counts have decreased by 62% over the past 50 years. Thanks to their poisons, cancer rates have increased from one in 20 people in the early 20th century to one in two people today.

To fight these people from military intelligence agencies, we must understand that most of the world's so-called politicians are just blackmailed and controlled slaves, or they are avatars of murdered politicians.

The Cabal Deep State has been talking for many years about killing 90% of the population on Earth and they have done their best. When I told this to friends and other people, they all thought I was crazy.

Many people are now aware and know what goes on behind the scenes. Will the rest of humanity wake up NOW?

Those in power want humanity to be caged. If people do not begin to recognize that power lies within our inner SELF, the forces of evil will continue to bring the masses under their spell, making them believe they have full authority over them.

Most of what they do consists of illegal actions, money laundering, fraud, mind control and manipulation. This type of monopoly is a system of central control that claims all power over billions of people.

Their list of other crimes is so long that only a Nuremberg-style war crimes tribunal can bring it all to light.

⸺◦⟨◦◈◦⟩◦⸺

Corey's dig put it in the right words on her blog *August 26, 2021*

what the World Economic Forum predicts for the world in 2030:
Normalization of QR Codes To Access Your Data, Your DNA, and Your BODY
https://www.coreysdigs.com/technology/the-global-landscape-on-vaccine-id-passports-part-4-blockchained/

The name of the game is to get all human beings and every product onto the blockchain for full traceability, where privacy will no longer exist. Think of the QR code as a middleman and the smartphone as a tool. Blockchain technology can be confusing, so here are some basics on it as well as how tokenization works. Originally created for digital currency transactions, it is now being used for IoT, supply chain tracking, financial services, assets management, identity verification, "smart contracts" and much more. A distributed ledger blockchain framework allows for the collection of data that is shared and synchronized across multiple sites, countries or institutions.

The QR code was never about a free donut or an easier way for people to shop or market products – that was just done to normalize its use and play it off as a "convenience," just like handy smartphones. The reality is that it's about controlling the human race by aggregating all data on every human being and object while allowing them full surveillance over your life, and scientists full access into your body. So the next time a restaurant provides a QR code to access their menu, demand an actual menu or leave the restaurant. Stop using the QR codes everywhere you go. Stop swiping your smartphone and playing right into their hands. REFUSE QR CODES…

Most people are familiar with Bitcoin operating on a "Decentralized" blockchain. But don't worry – the Linux Foundation is covering the "privacy" end of things, as covered in part 3, for all of these new blockchain platforms. The programs and agendas being developed and carried out for full control over humanity, involve **Federated, Centralized, and Decentralized systems**. There is an excellent chart that breaks this down on page 13 of the World Economic Forum's 2019 report on 'Trustworthy Verification of Digital Identities.'…Their ultimate goal is to

73

have every human being and every product traced, tracked, and surveilled. For example, the **Evrythng Product Cloud gives products a digital identity**. It creates a twin in the cloud which is linked to an identifier that is embedded into the smart packaging or smart code, making the object interactive with software intelligence that allows it to participate in new applications.

https://id4d.worldbank.org/guide/tokenization

https://101blockchains.com/blockchain-technology-explained/

https://www.youtube.com/watch?v=IBSLnaXbuUU

https://www3.weforum.org/docs/WEF_Trustworthy_Verification_of_Digital_Identities_2019.pdf

CHAPTER 7

PEDOPHILIA

When I was little my mother used to be very protective of me and my brother. In fact, when we went out in public she wanted us to be very close to her body at all times.

I remember that she would say to us that there were bad people that stole children, and their parents never saw them again. She told this to us hundreds of times, especially if we were going to be in a crowded place.

Between 1974 until 1979 my three children were born in USA. As they were growing up I was buying large amounts of milk at the supermarket, to feed it to the kids. There were pictures of children on those milk containers; all of them were children that had disappeared from families all over America.

Every time I bought milk there were different faces of children on the container. All of them had been lost and never found again. I was very puzzled... I was asking myself how it could be possible that so many children disappear daily in America?

Was my mother right? She knew all the way back then, in the 1950s and even before, about children disappearing in Italy?

When the pedophilia in the Catholic Church exploded in my face: I kept on wondering how it was possible that all those priests were involved in all of those horrible practices with children??!

My faith in the Catholic Church started to decline at that time. As I realized that we had been following what they told us since we were children and we obeyed them like sheep been taken to slaughter up until now. Is that called brainwashing…?

The wonderful Father Walsh had died at the age of 55. I missed him, as he was upright and I would have liked to ask him how all of that pedophilia in the Catolich church was possible.

I got to the point I stopped going to church.
Now all of the "stuff accumulated in my conscious mind" was starting to connect more every day.

This particular information will be extremely difficult for many.

The big blow came when I found out that there were humans without a soul on Earth who had to be drinking children's blood, raping them to harvest their soul power and eating their flesh, so they could live longer lives on Earth. When the body that they were using began decaying, they would jump in another body and they continued doing that for thousands of years!!!

I couldn't sleep any more at night for a long time. The pain in my heart for those children was too much to bear.

———◁•◇◈◇▷•———

"You may live to see man-made horrors beyond comprehension".   ~ Nikola Tesla~

In the month of August 2021 I found out that 800,000 children disappeared from the island of Haiti after the earthquake happened. Epstein had a submarine and with the help of the Clinton foundation he was rounding up the children of the island and transporting them... to who knows where!!!

"For we wrestle not against flesh and blood, but against the rules of darkness of this world, against spiritual wickedness in high places."

~~Ephesians 6:12~~

November 30--2022 From Benjamin Fulford

For months we knew about the indictments. Growing from 25.000 to 50.000 and now rising to 200.000. For years I have known about the many political, business and entertainment elites who like to hobnob with child sex traffickers like Jeffry Epstein and are involved in pedophilia.

As December 5--2022, there are 182,771 sealed, 21,381 unsealed and 409,162 non sealed indictments for child pedophilia and trafficking offences.

January 30, 2023 From Benjamin Fulford

The US government has been subsidizing child trafficking in all 50 states. This is thanks to the "adoption and safe families act," signed into law in 1997 by Bill and Hillary Clinton Rockefeller. Over $80 billion a year is being spent by the Federal Government for this program to give states over one million dollars for every child they seize from a family. The government itself admits that 83% of the children seized are done so under false pretenses.

It also now turns out they have been using fake drug tests to make it seem parents were "drug addicts" who had to have their children seized.

——————

Not one of my grown up children, friends, or relatives wanted to hear this news that I was unveiling, and I started to close myself up, even though I wanted to scream this truth with a megaphone. I wanted to tell America and the rest of the world that we need to wake up to this horrifying reality and eliminate these horrible parasites, Luciferian Cabal and Kazarian Mafia.

Pedophilia in this world is affiliated with Cultism, Satanism, Luciferianism, and Spirit cooking which is taking the soul of a raped child to gain Satanic Energy. This type of affair is at the higher levels of Governments. (Bill and Hilary Clinton, George Soros, elite rulers of the Earth, Kazarian mafia etc.)...

Child abuse is extensive and goes all the way back to the Vatican, for thousands of years.

In 2018, 10,000 Catholic priests were arrested in Australia for pedophilia, 750 pedophiles were arrested in California, another group of 400 arrested in Philadelphia, PA, at the end of May 2018, just to mention a few. I don't want to speak about Germany... This problem and ring of Satanism is extensive throughout the Globe.

There is and was a huge blackmail in the US Government officials, and other world governments, by the Kazarian Mafia, Luciferian Cabal. But because the Media lies, we do not know the total truth yet.

Children have been drugged and filmed at pedophiles' parties involving blood and animal dismemberment. The upper echelons, executives, VIPs of development agents and international bankers fund these parties.

Satanism goes on in world Banking and Corporations and also into Social Services.
That is why so many children are stolen from loving parents.
Politicians and those called "elite" are invited to these parties and they are secretly filmed and eventually become puppets of the Luciferian Cabal. When they have a film of a person, the Cabal grows in power and so many political decisions are made all over the world as a result of compromised politicians doing what they are told to do or else.

This is the way the world has been running for a long time. This increasing totalitarian control is attempting to take over the globe.

The individuals become handicapped by coming face to face with a conspiracy so monstrous, he cannot believe it exists. The American mind simply has not come to the realization of the evil, which has been introduced into our midst. It rejects even the assumption that human creatures could espouse a philosophy, which must ultimately destroy all that is good and decent.

Stop watching the mainstream Media, do yourself a favor, do your own research.

∞◄◆►∞

November 5

, 2022 Telegram Q the Storm Rider / Official Page

Jimmy Chèrizier also known as "Baberkyow " is a Haitian gang leader and former police officer who is the head of G9 (or G9 family and Allies) a federation of over a dozen Haitian gangs based in Port-an-Prince. In the late summer of 2022 Chèrizier took control of D.U.M.B.s for the Haitian. The D.U.M.B.S were used by the Clintons and Obama's black military projects operations. The underground base were the centering points for MASS human/children trafficking, Biolabs adrenochrome production through the Caribbean sea that supplied the Epstein, Maxwell, Rothschilds regime.

To this day hundreds of thousands of kids are missing in Haiti. The missing children of Haiti are openly connected to the investigations where children disappear by the thousands from hospitals every year.

Haiti was exposed to (DS) mil Projects that caused major Earthquakes and floods due to Tectonic weapons... This was intentionally ordered by the Rothschilds (DS inside of Pentagon) on behalf of the Clintons who were going to be exposed for human trafficking just weeks before ///
As Haiti collapsed inside the Earthquakes; the Clintons stole 80% of the Gold reserve in Haiti and overthrew the government in the aftermath of the color revolution.

The D.U.M.B.s in Haiti were connected directly to Epstein Island with 3 submarines occupying the base. The largest children trafficking operation ring of the Caribbean sea is centered in Haiti, with Bahamas, Dominican Republic, Jamaica, adults and children trafficking operations connected directly to Haiti labs D.U.M.B.s facilities ...just a stone throw away is Little Saint James Island ...Also known as Epstein Island.

Today Jimmy Chèrizier is fighting the Deep State and it is crucified by the world Media, world leaders as the exposure of Child trafficking/adrenochrome/BIOLABS/ inches closer to EXPOSURE as the U.S. White Hats get ready to take public control (overt) of the United State Mil. Operations connected to the Cabal crime families. OBAMA, CLINTONS, ROTHSCHILDS who are protected by the Blackhats of the FBI. PENTAGON, CIA, DOJ. Etc....

Haiti is inside a Military Attack that is about to happen. As Biden had sent his secretary of State Blinken to Canada and tried to get U.N. support to invade Haiti.

For many reasons Obama, Clintons, Biden, want to control Haiti because of the oil Reserve that Jimmy Chèrizier is blocking from the U.S. to the D.S. operations, the DS. U.S. Leaders have involved Haiti.

The U.N. (good White Hats infiltrate inside U.N.) have denied an invasion to Haiti and so CIA. DS, PENTAGON, CLINTON, BIDEN, Have turned to Bahamas for military invasion.
Haiti, this small nation is trying so hard to EXPOSE the U.S./ CLINTON FOUNDATION / U.N. and the disappearance of hundreds of thousand children in the past 10 years alone...Many go missing directly from Hospitals and Orphanages.

The DEEP STATE IS TRYING HARD TO COVER-UP the Atrocities they have created in Haiti. As Russia is in Communication with HAITIAN MILITARY COMMANDERS who desire to create a MILITARY COUP... THE U.S. is also aware of this Coup, and sent DS. Military intervention. In order to try to stop a FULL MILITARY COUP.

Here is a view of both sides of operations: White-Hats vs. Black-Hats:
For some reasons Black-Hats want chaos, climate change and confusion inside of the Diesel Collapse.
White-hats want exposure of a broken corrupt U.S. system, that is being guided by Democratic Party and the Deep State President that is taking command from NATO, U.N, Forces, DAVOS,

Rothschild... but one more important thing that the WHITE-HATS want is Justice for JFK and Human Trafficking.

This connection to U.S. shipping industry, connected to the U.S. mafia who controls the UNIONS (think of Jimmy Hoffa...how created the Unions and how he was killed for the UNION CONTROL and how the American Mafia was connected to the assassination of President FJ KENNEDY)/...

The Infiltration of: FBI. DOJ. The CIA, Democratic Party, into the U.S. Italian Mafia crime families was long ago established. (Mich like the Black Lives Matter and Antifa operations, are controlled by the Deep State...But also the heads of crime families of the Mafia are also controlled by Higher Levels...that are connected to UNIONS)///

The Mafia Heads DEEP STATE CABAL CONTROL THE U.S. Shipping industry that uses TRUCKING for GAS, DIESEL and for other operations, also control the Human modern days Human Slavery...

That's how you get over 840,000 missing kids a year in U.S. alone, with close to near 1.5 million Missing adults and children all together annually.

The ongoing current within close indictments are on the heels of operations of the Italian Mafia and their adherence to managing human trafficking routes through their Unions are near the END.

The long Journey of JFK JR. to destroy the US Syndicated Mafia families that are controlled by DS. OPERATIONS are in parallel jeopardy.

Most people think the Mafia is non-existent these days...and most other people don't believe in the Obama syndicate crime family. Or Clinton crime family.///

The operation to shut down the U.S. shipping system to near Holt is an EVENT that will restructure the shipping Regimen and take control from the elites, who control the PG 42 UNIONS and give the TRUCKING operations back to the people in the fall of 2024 into 2025

———◁◈▷———

November 6, 2022 The Storm Rider-Official Page

Putin was the first president to expose the WESTERN SATANIC OCCULT to the world on a wider scale... Even President Donald Trumps is now openly referring to Satanism as
" Horrific Satanic Rituals, happening in the U.S."
World known Lara Logan is also now coming out to claim things like
"Biden is trafficking kids so people can drink their blood."
From Hollywood elites Mel Gibson talking about human trafficking, blood sacrifices and blood drinking currently that control Hollywood and elites.

The list of satanic rituals involving Elites throughout the world is known and is being exposed daily, monthly and yearly since Q drop mass...Great Awakening Operation.
It has become a clear and present danger as Saudi Royal Bin Salomon has executed his own family for being part of Clinton's adrenochrome networks...in 2017.

The dark Generals in the US military who are loyal to Moloch God and Rothschild, Obama, Davos, Kazarian Mafia, know they are in trouble as the White Hats Military have all the evidence of human trafficking networks through the several US military bases abroad.
Time is passing, as LIGHT shines in the Darkest Parts of the Satanic Cabal Agendas and Regime.

Those who control the money control the world, but the world is changing, because the majority of humanity is awakening, everything is starting to change.

The media on TV doesn't tell you!
Right now in these days of chaos, everything is backwards:
Most of the doctors kill health.
Lawyers destroy justice.
The Government destroys liberty, freedom of speech and freedom of choice.
Religion destroys spirituality.

The reality of Child trafficking, human trafficking and Satanism are things that were triggered by a WikiLeaks data dump of emails that happened in the weeks before the election of President Trump.

In the future weeks and months so much more monumental news will come out.

"For nothing is hidden that will not be made manifest, nor is anything secret that will not be known and come to Light!"

~~Luke 8:17~~

This new article by Corey Digs opened my eyes even more than before about what is going on in the school system. Every parent and family members that care about their own children must read this peace about:

By Corey digs: Indoctrination Programs (in the schools)

There is much more on this site on this link.
https://www.coreysdigs.com/health-science/funding-the-control-grid-part-2-the-psychological-framework/

- According to a presentation hosted by Parents Rights in Education, in 2010 the Consolidated Appropriations Act signed by Obama sent federal funding through HHS to the CDC's Department of Adolescent and School Health (DASH), which created a partnership with the Department of Education on a series of grants to accomplish three goals for school-based programs across the nation: to implement comprehensive sexual education, to create school-based access for sexual health services, and to ensure "safe, supportive environments," in which to implement these programs. As the presenter explained, the CDC's DASH playbook in layman's terms means "They're going to teach it. They're going to give the kids access to it. And they are not going to tell the parents anything about it or let them do anything about it." By 2018, federal grants were distributed for student access, including minors, to school-based sexual health services including birth control pills, condoms, IUDs, hormonal implants, emergency contraception, and referrals to Planned Parenthood. As part of California's Behavioral Health Initiative, in 2019 Planned Parenthood announced that they would be opening 50 on-campus clinics in Los Angeles high schools. At least 5 Planned Parenthood clinics are currently operational in high schools in the Los Angeles area.

- Planned Parenthood has been in the "gender transition business," since 2017 and has shifted much more focus towards a "gender affirming" revenue stream since then. With 239 clinics in more than 40 states offering transgender services, Planned Parenthood has made it quite easy for thousands of kids to access hormone treatments. In some states, "gender affirming" care may be provided to minors without parental consent or knowledge. Furthermore, public schools continue to use taxpayer dollars to hire Planned Parenthood or their affiliates to teach sexual education programs including the promotion of abortion, gender fluidity and transitioning. Each year, 1.2 million students receive sex education from Planned Parenthood's affiliates.

- And more, much more...

——◦◦◦——

When I searched online to know what HHS was about, I got his answer:
The **mission of the U.S. The Department of Health and Human Services (HHS) is to enhance the health and well-being of all Americans**. (what a name !!! Is it a joke? To me they are really a nefarious bunch, it reminds me of the **spiritual wickedness in high places**, which was spoken in the book of revelation).

——◦◦◦——

The DEFENDER 3/16/2023

Report Linking Fluoride to Lower IQ in Children Made Public After CDC, HHS Tried to Block It.

BY Brenda Baletti, Ph.D.

The National Toxicology Program (NTP) on Wednesday released a draft report linking prenatal and children fluoride exposure to reduce IQ in children, after public health officials tried for almost a year to block its publication.

The U.S. Department of Health and Human Services (HHS) and the Centers for Disease Control and Prevention (CCD) initially blocked the NTP from releasing the report, according to emails obtained via a Freedom of Information Act (FOIA) request.

But a court order stemming from a lawsuit filed by Food and Water Watch against the U.S. Environmental Protection Agency (EPA) forced the report's release this week.

The NTP, an interagency program run by HHS that researches and reports on environmental toxins, conducted a six-year systematic review to assess scientific studies on fluoride exposure and potential neurodevelopmental and cognitive health effects in humans.

The report, containing a monograph and meta-analysis, went through two rounds of peer review by the National Academies of Sciences, Engineering, and Medicine. Comments from reviewers and HHS and NTP"s responses also were included in the report released Wednesday.

According to its website, the NTP "removed the Hazardous classification of fluoride" in response to comments in the peer-review process. Yet, the report states:

"Our meta-analysis confirms results of previous meta-analyses and extends them by including newer, more precise studies with individual-level exposure measures.

THE DATA SUPPORT A CONSISTENT INVERSE ASSOCIATION BETWEEN FLUORIDE EXPOSURE AND CHILDREN IQ...

The results were robust to stratifications by risk of bias, age group, outcome assessment, study location, exposure timing, (including both drinking water and urinary fluoride)". " These findings fly in the face of the empty, unscientific claims U.S. health officials have propagated for years, namely that water fluoridation is safe and beneficial," said Robert F. Kennedy, Jr., Children Health Defense chairman and chief litigation counsel. " It's past time to eliminate this neurotoxin from our water supply."

The controversial report will play a key role in determining the outcome of a lawsuit brought in 2017 by several nonprofits against the EPA to end fluoridation of drinking water, plaintiffs' attorney Michael told the Defender.

We had to fight hard to have this report even made public," Connect said. "They [CCD and HHS] buried this. If they had gotten their way, this report would have never even seen the light of dai." Connect said.

Since the trial began in 2020, U.S. District Judge Edward Chen has been waiting for the NTP to complete a systematic review of fluoride neurotoxicity before ruling on the case.

Groups like the American Dental Association publicly pressured the NTP to "exclude any neurotoxin claims" from the reports.

Connett said during the trial, the EPA repeatedly claimed that the plaintiffs' allegations about toxicity could not be verified because there was no "systematic review."

The document released Wednesday fill that gap.

Connett said:

"So now what do we have? We have a systematic review by one of the pioneering, leading, most authoritative research groups on toxicology in the world."

"They just completed a systematic review that took them six years to complete, so if that's not enough to demonstrate a hazard under the toxic substances control act, then how would any citizen group ever be able to meet the standard?"

https://childrenshealthdefense.org/defender/ntp-report-fluoride-lower-iq-children/?utm_source=salsa&eType=EmailBlastContent&eId=1faf87d9-06b7-4194-8160-33245d49b37a

Coming Jan. 13– 2024 : 'Fluoride on Trial' Documentary Exposes 70 Years of Censored Science

In "fluoride on Trial: The Censored Science on Flouride and your health" a new documentary aired an CHD.TV Saturday jan.13 attorney Michael Connett and Children's Health Defense's Mary Holland expose the long history of Government and Industry suppression of scientific research revealing the toxic effects of fluoride, particularly on children.

⊶⊱⊰⊷

In Italy while I was growing up, we never had fluoride in the water, until recent times. Our water was coming straight from the glaciers of the Appennini Mountains. As a child I always had a big imagination. When I came to live in the U.S., some people told me that the American water was not good, I didn't like the taste of it and I drank very little water.

About 40 or more years ago I learned that fluoride calcifies the Pineal Gland not only lowers the IQ in children. I have used filtering systems in my kitchen and changed it over the years as the filtering system was improved. Then I found the reverse osmosis system from Dr Mercola.

This is what the National Institute of Health (GOV) tells us:
 "The main function of the pineal gland is to receive information about the state of the Light-Dark cycle from the environment and convey this information by the production and secretion of the hormone melatonin".

This is what I have known for a long time: The Pineal gland is also known as AJNA or Third Eye. It is located in the center of the brain in between the eyebrows. It is connected to the Light and our Soul. It looks like a little pine cone. It helps us to imagine things with our mind's eye before we create them in a physical form. This gland is most heavily targeted and poisoned by the deep state, because it serves to us as a connection between the physical and the spiritual world. According to our Ancestors and their teaching, it is the key to higher states of consciousness.
In ancient Egypt, the pineal gland was known as the seat of the spirit of the Soul.
Jesus said: the eye is the lamp of the Body and if the eye is clear, the body will be filled with Light.
This is related to the eye of intuition that makes us see things clearly and in truth.

That is the reason why the Deep State and KM wants our pineal gland to be calcified by putting fluoride in our drinking water, so our imagination and mind's eye cease to function.

They have been doing this evil for many years because they do not want us to be able to imagine and create. They want us to see only the reality that they, the wicked ones, create for us, on TV and cell phones, which is practically just lies, deception and illusions.

"WHEN IS AMERICA GOING TO WAKE UP TO THE FACT THAT THEIR GOVERNMENT IS NOT CARING FOR ITS PEOPLE? IT IS ACTUALLY AGAINST THE PEOPLE BECAUSE THE GOVERNMENT HAS BEING INFILTRATED AND HELD HOSTAGE BY A FOREIGN GROUP OF PSYCHOPATHS CRIMINALS THAT WANT TO OWN US AS SLAVES AND FOR THE MOST PART, WANTS US STUPID OR DEAD."

CHAPTER 8

KHAZARIAN MAFIA

I followed David Wilcock for many years. I read his first book The Source Field Investigation as soon as it was published in 2012, because he was talking about some of the things that interest me, other things I had learned over the years, and others I learned from him. Eventually I read all the other books he wrote.

I first heard the name Khazarian Mafia many years ago from Benjamin Fulford in a talk he had with David Wilcock on YouTube.

Benjamin Fulford, Canadian born, was ex-chief editor of Forbes in Japan where he lives. Since I found out the name Kazarian Mafia, I've always wanted to know who they were, where they resided and how bad they were.

Being Italian I always knew about the Sicilian Mafia, and the Mafia of Calabria, which is another Italian mafia called "Ndrangheta." They are a dark evil group of individuals.

As time went by I learned more about this horrible Khazarian Mafia and I found out that they drink human blood. That made me immediately think that they were beings that were assimilated by the demonic Archons.

The 13 families that appointed themselves as the rulers at the top of this world are the Khazarian Mafia, which are possessed by the Archons. They interbreed and will not marry further down than second cousins. They drink human blood. They believe they are different from the rest of the human population. They believe that their DNA is different from our DNA and they have a core belief of superiority over all other humans.

They want to get rid of the white human population because they consider them to be uncontrollable and combative, not too easy to rule. The entire white race of humans is fighting the Khazarian Mafia whether they know it or not.

The Khazarian Mafia is also called the Ukrainian Nazis. They believe that being selected and carrying a host (demon/archons) is a good thing and makes them special.
These are the people that the Russians have been fighting against since the year 759 AD because they knew at that time that they drank human blood.

The Khazarian Mafia keeps a lot of secrets and only reveals them to their initiated members. This is the same model they used for all their creations: Freemasons and all other secret societies.

It is the same model they used to create the Catholic Church/Christianity which was infiltrated by the Khazarian Mafia during the Byzantine Empire.

I believe though that it was even before that time because I believe they came up from northern Africa, into Rome. But this is only my opinion.

Now the Russians are going to start releasing the evidence of "demon investigations, including" where the Khazarian were using live humans in experiments, This is called "demonic possession."

It is the Khazarian that from time immemorial have created 10,000 laws from the original 10, in order that they may pull the wool over their brother's eye when they felt the need to do so.

The antiChrist Khazarian Mafia, Baal is not just one individual but many, whose beliefs and actions work against the Christ Consciousness of Love and inner-standing awareness.
As their dishonest narrative collapses around them, the mainstream media outlets and the corrupted politicians are trying together their latest bit of distraction to keep the narrative of lies and fear going in other directions than the truth.

All the pandemics in history were created by them and some are: the Black Plague, Leprosy, Spanish Flu, the Bird Flu, the Swine Flu, Mad Cow Disease, Lyme Disease, AIDS and the list is probably much longer than this, because they created illnesses every time the population on Earth grew to large in numbers. They needed to keep humanity down in numbers to be able to control it. And that is what they are doing now with Covid19/ every 6 months new infections

.

—⊃·≺◆◇◆≻·⊂—

"Infertility" A Diabolic Agenda. (Infertilitymovie.org).

In the weekly Geo-political News and Analysis by Benjamin Fulford article of May 16, 2022, here is the link:
https://benjaminfulford.net/2022/05/16/destroy-km-or-face-nuclear-war-russia-and-china-tell-pentagon/

I copied part of the article, shown below:

Destroy Khazarian Mafia or face nuclear war, Russia and China tell Pentagon

1- The Russian delegate handed over documents and evidence in the session record confirming the following:

   a) Official Pentagon funding for an apparent biological weapons program in Ukraine

   b) Names of American people and companies specialized in the evidence and documents involved in this program.

   c) The locations of laboratories in Ukraine and the attempts made so far to hide any evidence

2- The locations of the American laboratories that manufacture and test biological weapons in 36 countries around the world (12 countries more than the previous session).

3- The diseases and epidemics, the means of their release, the countries in which they are being tested, and when and where the experiments were carried out with or without the knowledge of the governments of these countries.

4- Among the experiments and relics is the virus responsible for the current pandemic and the huge number of bats used to transmit this virus.

5- The peoples of America, France and Britain have, under the psychological pressure, been brainwashed into believing a fictional version of what is happening.

6- The World Health Organization denies knowledge of the biological experiments in Ukraine despite proof provided by Russia that its representatives regularly visited and were in correspondence with suspicious American laboratories around the world.

7- China is asking why the US and UN desperately refuse to conduct an investigation by specialists to find out the truth, especially given the documents and compelling evidence?

The Russians captured a large number of migratory birds that had been numbered and had microchip-controlled capsules containing various plagues attached to their bodies, Russian and Chinese sources add. When these birds fly to Africa, Asia, Latin America etc., the plague vials could be opened by satellite command where they would cause the most damage.

Russians also accused President Joe Biden, former President Barack Obama, former Secretary of State Hillary Clinton, and billionaire George Soros of involvement in the conspiracy. The Chinese and Russians insist that "whoever commits such immoral and inhuman acts must be punished."

⸺◉⟨⬦⟩◉⸺

On May 30, 2022 Benjamin Fulford said:

"The end is near for the fake Biden regime as Jeff Bezos, one of its chief bankers, has bailed out and joined the Earth alliance, CIA sources say…The Ukraine government and military are collapsing. The Khazarian Mafia project to rule the world from a greater Kazaria (Ukraine + Kazakhstan) is thus now totally doomed."

The other thing to notice is the complete collapse of Davos World Economic Forum and his sister organization the World Harm Organization (WHO) as they gathered in Switzerland in June 2022

The WHO was supposed to mark its 75th anniversary with a victory lap granting its head absolute dictatorial powers over the world governments. Instead its top staff has been arrested for spreading bio-weapons, Mossad sources say.

Here is a piece of history that was new to me:

President Abe Lincoln and Kazarian Mafia-- By Clif High
https://www.disclosurenews.it/abe-lincoln-and-khazarian-mafia-clif-high/

Deep State, Cabal, Illuminati and the Khazarian Mafia are all the same group and have been actively working toward mass depopulation.

Now we realize that this isn't about what's right for us. We are now in a real war with the forces of evil that want to change the mentality of the masses, and **the transhumanism agenda replacement program (**depopulation of the world). It is about raw nakedness and absolute power at all cost.

We must unlearn what we have learned in the school systems and in religions. We have the right to learn the truth and we have the right to be free and prosper, and it is up to us to stop this evil.

"Never change things by fitting the existing reality, to change something, build a new model that makes the existing model absolute!"

~~R. Buckminster Fuller~~
(1895-1983)

We are in a great war …Good vs. Evil. The time is nearly over for the Baal and his devils, even though it is not going down easily.

The awakened ones are seeing the internal conflict of those souls that are still going to church asking for help from a God outside of themselves, instead of going into their inner self. They don't realize that evil sucks all their energy out, right there in Church. They still believe that there is a God/Government that will fix things for them.

People do not understand that we are the children of the LIGHT, We are the INCARNATED VESSELS OF THE SOURCE OF ALL THINGS. We are powerful and we can change the world around us if we use our powerful Spirit/Mind to manifest peace, prosperity, love and get rid of the control of the few.

The Book of Revelations tells us that we have a lot of help from the higher realms. Without their help the entire Earth and population would be gone. The Enlightened Beings from higher dimensions have the desire to save as many earthlings as possible to transition to 4th and 5th dimension, but they still face serious difficulties.

To accelerate the destruction of the remnant dark energy of Atlantis the co-creators and Higher Forces are reactivating all the 144,000 groups of pyramids to help Mother Earth. Our planet cannot cope any more with the legacy of Black Archons and their Draco's –Reptilians.

The pyramids bring positive energy from other Stars in the universe to the Earth's core!

Let's start taking control of healing our physical body and planet Earth. and get ready for cosmic ascension by removing the toxins, fear, pain, and loss. Let's learn to forgive and detach from our personal old stories and trauma.

We must overcome psychic warfare, demonic forces and mind control, because evil doesn't make spiritual beings, it makes brainwashed puppets.

Things on Earth are moving faster and faster. That's why it seems to us that the 24 hour day is shorter because the frequency of the Earth is higher and our bodies are changing accordingly.

The passage of time has literally accelerated the pace, even as we still try to regulate and organize ourselves for our daily activities; the day is rushing through too fast.

Time seems to have halved.

* One year is like 6 months
* One month is like 2 weeks
* A week is like 4 days.
* One day is like 12 hours.
* One hour is 30 minutes.
* One minute is like 30 seconds.

In the Holy Bible there is this statement "If these days are not shortened, no flesh will survive."

Abandon all Earth based religions which were invented by the Archons to keep us separated ("my God is better than yours!") Religions created all the wars in history. Abandon the belief systems and industries that dominate and control the divine feminine power in our world and create an energetic imbalance that negatively affects every man, woman and child on the planet. The patriarchal model, enforced by the Khazarian Mafia, is obsolete.

We must reconnect with our Higher Self to restore balance, and we must develop a positive mind set now.

We are going through a massive shift of the Ages. Those working against humanity and consciousness expansion are using all their dirty work to hold the denial of the soul's purpose which is to achieve higher consciousness and perfect peace.

Their evil ways allow loosh to be generated and extracted for satanic agendas. Loosh is a term applied to the energy produced by humans and animals that other entities use to feed from.

——◦‹◦†◦›◦——

"This 3D world has been socially engineered by these entities to produce a myopic mental polarization by gratifying materialistic excessive pursuits, to produce a spiritually bankrupt population.

When there is no value or meaning to life, there is no accountability, no morals or ethical consideration toward the consequences of actions that are directly related to increased pain and human suffering in the world. World War I and II are examples we should have learned from."

~~Lisa Renee~~

Things probably are going to get worse before they will get better because the world's sleeping sheep must see the real darkness in the world. It has to be this way in order to wake up the majority of humans to see the satanic system that has been in control of planet Earth for millennia.

The real story is coming out, the Deep State, Cabal, Illuminati, Khazarian Mafia, the 13 families that want to continue to control the entire Planet and Humanity. They are losing and will not be able to hide the evil they have perpetuated over thousands of years.

The Deep State Khazarian Mafia is now in defense mode and this is a sign of weakness. The hunters are being hunted now.

We need to be hopeful because this is the sign of a great change for the Earth and Humanity. This great change will give us a consciousness renaissance.

We need to focus our attention on the maintenance of our form in a state of purification. That way we will maximize our chances to complete the process of accessing consciousness to higher levels of density.

We must realize the power we hold by being connected to the ETERNAL SOURCE of all things, the LIGHT IS GOD and we are getting showered with an immense amount of LIGHT WAVES from the Central Sun of the Galaxy these days. The dark evil cannot take this amount of light, they live in darkness, and their power is weakening.

——◦‹◦†◦›◦——

More news from Benjamin Fulford's weekly report, May 23, 2022 titles:

Top Secret negotiations for New Age Proceed Well as Western rule Collapses.

Top-level negotiations between Asian and Western elders to start a new golden age for humanity are proceeding well, according to sources involved. The basic agreement calls for the complete write-off of all debts, public and private, a one-time redistribution of assets and a massive campaign to end poverty, stop environmental destruction and colonize the Universe with Earth life. This plan is supported by—among others—the Western committee of 300, the Russian government, the Indian government and the Asian secret societies that control China, Asean, Korea and Japan. There are some concrete moves involving massive amounts of off ledger gold and dollars taking place. The details cannot be publicly disclosed for security reasons, the sources involved in the negotiations say.

Of course, we have heard this talk before, so believe it when you see it.

For now, the task is to make sure the existing Western ruling structure collapses without destroying the planet. Already, the Western ruling class has fallen into what can only be described as collective insanity as their control grid collapses. The latest sign is a massive campaign of monkeypox fear porn. The fear porn appears to be a desperate attempt to justify a massive power grab by the WHO and the Davos, World Economic Forum.

As an example, last week the Biden avatar called a summit of the "Americas" only to be boycotted by the majority of states in the region. Cuba, Venezuela, Nicaragua, Mexico, Guatemala, Honduras, El Salvador, Uruguay and Bolivia were among the countries to boycott the meeting. Leaders of countries that did attend, like Argentina and Belize, took the opportunity to criticize "Biden."

The rest of the world, including Africa, the Middle East, Asia (except for the Japanese and South Korean slave states), and a large part of Europe is also shunning the openly criminal US Corporate government. That is why an international boycott against it is causing an economic free-fall there.

On June 13, 2022 Benjamin Fulford added:

The world is going through a fundamental shift as the Western world is being cured of its satanic infection. The result will be an unprecedented era of world peace and prosperity. This will be the legacy of the "make love not war," hippy generation.

However, before this new age can start, some final housecleaning is being done. This can be seen in the international isolation and financial collapse of the fake Joe Biden regime in the United States. The collapse is accelerating as the Khazarian Mafia "Biden" regime falls into pariah status.

The Georgia GuideStones took 9 months to be erected in 1980, it was a monument of the New World Order. It was erected by the ruling elites so they could prolong their lives with the help of new technologies and with the intent to reduce the world population to 500 million people.

The megalithic structure was located in the "Granite Capital of the World" in Georgia. The first two stones indicated those who created it and why. This structure was erected on the surface of the flat top of the hill, but on the subtle plane, this monument was the location of the Global egregores of Black Archons and their dark Hierarchy.

Guiding Stones being removed

Here is a little peace of what was written on the stone:

"Maintaining humanity under 500.000.000 in perpetual balance with nature.
Guide reproduction wisely improves fitness and diversity,
Unite humans with a new language.
Rule passion-faith-tradition and all things with tempered reasons…"

These guys had decided to eliminate 90% of the population and keep the rest in slavery, and they wrote it in stone.

Well on July 6, 2022, their monument and altar were completely taken down. This was the most visible takedown of the Kazarian Mafia in the US.

According to Benjamin Fulford though, there was more visible news in our western world: the assassination of the Japanese Prime Minister Shinzo Abe (connected with KM), the messy removal of British Prime Minister Boris Johnson. Also the changes of the governments in Estonia, Ireland, Israel and Sri Lanka also fell.

The Dutch farmers completely shut down the country.

In France, Macron's party lost the parliamentary election and now the criminal investigation against Macron has started.

In the United States, the Joe Biden show has become so farcical that even the KM propaganda media outlets like the New York Times and Washington Post are starting to question the whole show.
In these hot summer days the archaic world leaders, who are focused on selfish interests of economic survival are going to bring the population into a more acute and explosive crisis that will materialize a complete reboot of the planet order and will shake the geopolitical, financial and economic system.

July energies are divided in two periods. The first one will last until July 26, 2022. This will fuel emotional tension and belligerence, especially in the countries with unstable governments.

Rothschilds and Rockefellers seek to surrender as Germany joins the planetary liberation alliance.

There is still a lot of work to do, but maybe we are starting to see a little light at the end of the tunnel.

Feb 6, 2023 From Benjamin Fulford

The Khazarian Mafia has been comprehensively defeated, mopping up continue

Events over the coming weeks will make it obvious even to the most brainwashed sheeple in the West that something fundamental and historic has taken place. There can be no doubt the Khazarian Mafia has been defeated and the mopping up of their last leaders has begun. The arrest of New Zealand cross-dressing Prime Minister Jack Adern, criminal charges against Swiss President Alain Berset and Pfizer, mass arrests of Ukrainian government officials and many other events all point to this.
Another sign the KM is losing is that it is now becoming mainstream to talk about Mossad and Israeli founding Prime Minister David Ben Gurions' involvement in the assassination of President John F Kennedy.

Israel meanwhile is trying to negotiate a surrender to the Russians via France. The bankrupt US Corporate Government, for its part, became the laughing stock of the world by using "a balloon ate my homework" excuse after Rockefeller slave Secretary of State Antony Blinken was told not to bother coming to China on a begging mission.

The war in Ukraine is also ending with the last Satanists now surrounded in Bahamut where they are expected to fight to the death over the coming weeks.

Most importantly of all, the KM lost control of the world's financial system on January 31st and the quantum financial system is now operating (more about that later).

However, it is not over until it is over and there is a strong possibility the desperate and cornered KM will try some sort of black swan mega-terror event, most likely either an EMP or nuclear attack.

Turkey's interior minister Suleyman Soylu told the US ambassador to take his "dirty hands off of Turkey" after Washington and eight European countries issued travel warnings over possible terror attacks there. Similar warnings were made about Iran.

The order by Western governments to evacuate their citizens means it's likely the cornered Israeli regime is launching a massive attack there. A 7.8 magnitude earthquake followed by dozens more as this report was being written indicates it has already begun. "This is a weather modification weapon being used to show Erdogan who is still in charge," a CIA source says. The attack took place as Russian Foreign Minister Sergei Lavrov landed in Iraq to launch "military-technical" cooperation with Turkey, Iran, Syria and Iraq along with the Gulf States. Since Egypt is also a de facto Russian ally this represents an existential threat to the KM regime in Israel (not to the Jews though).

The Iranian government has now warned Israel's neighbors that in order to "maintain their security," they should "stay away from the regime that is dying every day."
The KM is desperately trying to blow up the whole planet (using agents in both Iran and Israel) because the KM-controlled Vatican and their US Corporate Subsidiary lost control of the financial system on January 31st. That is why the entire Italian internet shut down on February 1st, MI6 sources say. The Vatican Bank, which was used to bribe most so-called world leaders, has been cut off. Now the KM have until February 13 before their US Corporation is completely cut off.

<hr/>

By <u>Benjamin Fulford Weekly Reports</u> November 7-2022

The Khazarian mafia stranglehold on the planet earth is finally ending. Germany and Japan have joined Russia, China and England in the planetary liberation alliance. This has forced the Rothschild and Rockefeller families hiding in Zug, Switzerland to contact the White Dragon Society to negotiate surrender.

However, until the United States is liberated from the Khazarian mafia, and other dark entities, the war will continue. The key is to prevent the theft of the mid-term US elections by the KM during the blood moon eclipse Election Day of November 8th. The US military White Hats promise to declare war on the KM if the election is stolen.

There are also undeclared wars raging in Brazil, Pakistan, Israel and elsewhere that will have to be settled. Then, of course, the war in Ukraine also needs to be finished. However, there can be no doubt planetary liberation is imminent.

Let us start with the situation in Germany. There Donatus, the Prince and Landgrave of Hesse, is leading Germany to independence for the first time since World War II. Donatus is a descendant of Queen Victoria, the German Emperor Frederick III, Caesar Victor Emmanuel III of Italy etc.

Donatus sent Chancellor Olaf Scholz to China last week to negotiate Germany's entrance into the alliance. The fact he reached a deal can be seen in reports in the official Chinese and Russian state news agencies.

According to sources, Xinhua Sholz told Chinese President Xi Jinping "a multi-polar world is needed"…Germany "opposes bloc confrontation" and supports "peace talks and to build a balanced, effective and sustainable security architecture in Europe." In other words, Germany is ready to dump NATO and replace it with something more inclusive.

CHAPTER 9

US MILITARY INDUSTRIAL COMPLEX AND MILITARY WHITE HATS RESISTANCE

William Astore is a retired lieutenant colonel of the U.S. Air Force, who has taught at the Air force Academy, the Naval Postgraduate school and taught history at the Pennsylvania College of Technology.
He posted an article on the DEFENDER in this month of February 2023

"Children's Health defense News and views"
https://childrenshealthdefense.org/defender/united-states-military-industrial-complex/?utm_source=salsa&eType=EmailBlastContent&eId=71b08967-8feb-40be-adf9-2937efc0092c

this below is only a part of it:

US MILITARY INDUSTRIAL COMPLEX IS CHOKING DEMOCRACY, HOW DO WE STOP IT?

America's founders were profoundly skeptical of large militaries, of entangling alliances with foreign powers and of permanent wars, according to Bill Astore, a card-carrying member" of the military-industrial complex, who warns: "So should we all be."

My name is Bill Astore and I'm a card-carrying member of the military-industriaL complex, or MIC. Sure, I hung up my military uniform for the last time in 2005. Since 2007, I've been writing articles for TomDispatch focused largely on critiquing that same MIC and America's permanent war economy.

I've written against this country's wasteful and unwise wars in Iraq and Afghanistan, its costly and disastrous weapons systems and its undemocratic embrace of warriors and militarism. Nevertheless, I remain a lieutenant colonel, if a retired one. I still have my military ID card, if only to get on bases, and I still tend to say "we" when I talk about my fellow soldiers, Marines, sailors and airmen (and our "guardians," too, now that we have a Space Force).

So, when I talk to organizations that are antiwar, that seek to downsize, dismantle, or otherwise weaken the MIC, I'm upfront about my military biases even as I add my own voice to their critiques. Of course, you don't have to be antiwar to be highly suspicious of the U.S. military. Senior leaders in "my" military have lied so often, whether in the Vietnam War era of the last century or in this one about "progress" in Iraq and Afghanistan, that you'd have to be asleep at the wheel or ignorant not to have suspected the official story.

Yet I also urge antiwar forces to see more than mendacity or malice in "our" military. It was retired general and then-President Dwight D. Eisenhower, after all, who first warned Americans of the profound dangers of the military-industrial complex in his 1961 farewell address.

Not enough Americans heeded Eisenhower's warning then and, judging by our near-constant state of warfare since that time, not to speak of our ever-ballooning "defense" budgets, very few have heeded his warning to this day. How to explain that? Well, give the MIC credit. Its tenacity has been amazing. You might compare it to an invasive weed, a parasitic cowbird (an image I've used before), or even a metastasizing cancer. As a weed, it's choking democracy; as a cowbird, it's gobbling up most of the "food" (at least half of the federal discretionary budget) with no end in sight; as a cancer, it continues to spread, weakening our individual freedoms and liberty. Call it what you will. The question is: How do we stop it?

What does matters to the Military Industrial Complex isn't either the truth or saving your taxpayer dollars, but keeping those weapons programs going and the money flowing. What matters, above all, is keeping America's economy on a permanent wartime footing both buying and less new (and old) weapons systems for the military and selling them globally in a bizarrely Orwellian pursuit of peace through war.

Indeed last year, Congress showed $45 billion more than Biden administration request (more even than the Pentagon asked for) to the MI Complex all ostensibly in your name. Who cares that it hasn't won a war of the faintest significance since 1945.

Even the "victory" in the Cold War (after the Soviet Union Imploded in 1991) was thrown away. And now the MI Complex warns us of an arousing "new cold war" to be waged, naturally, at tremendous cost to you, the American people taxpayers.

As citizens, we must be informed, willing and able to act. And that's precisely why the complex seeks to deny your knowledge, precisely why it seeks to isolate you from its actions in this world. So, it's up to you-to us- to remain alert and involved.

Most of all, each of us must struggle to keep our identity and autonomy as citizens, a rank higher than that of any general or admiral, for, as we need to be reminded, those troops wearing uniforms are supposed to serve us, not vice versa.

I know you hear otherwise. You 've been told repeatedly in these years that it's your job to support the troops. Yet, in truth, those troops should only exist to support and defend you, and of course the Constitution, the compact that binds all together as a Nation.

When misguided citizens genuflect before those troops (and ignore everything that's done in their name) I'm reminded again of the Elsenhower sage warning that only Americans can truly hurt this country.

Military services may be necessary, but it's not necessary ennobling.

American founders were very skeptical of large military alliances with foreign powers of permanent wars and threats. So should we all be.

Is Citizen United the answer? No, not the "Citizen United" not the case in which the Supreme court decided corporations had the same free speech rights as you and me, allowing them to co-opt the Legislative process to drown us out with massive amounts of "speech" aka dark-money-driven propaganda.
We need Citizen united against America's War machine.
Understanding how that war machine works, not just its waste and corruption, but also its positive attributes, is the best way to wrestle it down, to make it submit to the will of the people. Yet activists are sometimes ignorant of the most basic facts about "their" military.

Today March 30- 2023 I listened to a video:
The DEFENDER SHOW ROBERT F. KENNEDY, JR. WITH SASHA LATYPOVA

https://childrenshealthdefense.org/defender/military-covid-vaccines-rfk-jr-podcast/?utm_source=luminate&utm_medium=email&utm_campaign=defender&utm_id=20230330

As Kennedy says: This Speech is immensely important because it puts a new dimension on the corruption that I would call a coup d'état against democracy by forces that include not only the medical cartel and our military intelligence, but it's apparatus as well. This is a huge fraud they have pulled off that has taken off the entire regulatory community, physicians and the public.

The U.S. government's COVID-19 vaccination effort is a biological weapon project run by the U.S. Department of Defense (DOD), according to Alexandra Latypova, a former pharmaceutical research and development executive with 25 years of industry experience.

Latypova, who oversaw compliance for more than 60 clinical trials, knows the regulatory standards pharmaceutical companies historically were required to meet before bringing a product to market.

"People misunderstand that this is another instance of Big Pharma corruption," she told Robert F. Kennedy, Jr., chairman and chief litigation counsel for Children's Health Defense, during an episode of " R.F Kennedy Jr. The Defender Podcast." "It's much, much bigger than that."

Latypova said we have government reports describing the COVID-19 vaccines as biological weapon. "I have a question to our government," she said. "What is it that they're exactly forcing on us?"

The DOD is "fully in charge" of the COVID-19 vaccine's manufacturing and distribution, and it owns the vaccine "until it is injected into a person," she said.

By creating a "pseudo-legal structure" over time that included Emergency Use Authorization (EUA) and other transaction authority agreements--called OTAs--the U.S. government allowed the military to take over the distribution of vaccines without adhering to historical safety testing guidelines or product recall procedures.

According to Ladylove, the notion that the COVID-19 vaccine met regulatory standards for safety and effectiveness was the "biggest lie that was sold to the public."

"I am describing a very illegal structure that's made legal on paper," she said. "It's unlawful. They--the government--are driving this."

Kennedy agreed with Latypova and pointed out that OTA was designed to allow the Pentagon to quickly buy weapons and systems without paying attention to any existing regulatory authorities.

Kennedy said:

"What they've done is they've taken that authority and they've applied it to the vaccines so they're purchasing the vaccines under OTA as a demonstration product.

"It's all a huge military operation and the involvement of the drug companies is a kind of window dressing." "The DOD paid the pharmaceutical companies for their brand names so people would think they were getting something from Pfizer or Moderna-- but all the distribution and manufacturing is done by the military," Kennedy said. " the pharmaceutical companies were brought in to put their name on it and then to pretend to do clinical trials," he said.

Latypova and Kennedy discussed how the military accomplished this without most workers involved in the production and distribution of the vaccine catching on. They also discussed how citizens and lawyers might effectively challenge the Pentagon's COVID-19 vaccination project in the court system.

On February 25 -2023 article on Dr. Mercola Newsletter

In an interview with journalist Kim Iversen, Robert F. Kennedy Jr. explains how the military industrial complex uses mind control techniques and fear to exert control over the population.

Fear is the enemy, as it allows totalitarian systems to take control of people, destroying democracy in the process.

While democracy is resilient, we now have the technology available to control human behavior at a large scale.

Democracy is dependent on the free flow of information, while censorship leads to totalitarianism.

Robert F. Kennedy Jr. speaks the truth about the authoritarian pandemic response that continues to threaten democracy and liberty as we know it.

It cost him friendships and 40 years of political contacts, not to mention loss of income and business relationships.

But the threats to his reputation and credibility, as the media have attacked him and his message, don't feel like a sacrifice, Kennedy says, as he feels called upon to advocate for this issue.

In the interview with journalist Kim Iversen, Kennedy explains, "I look at it as a gift. I was raised in a milieu, in a family, where we assume that our lives would be consumed in some controversy, and that it would be a privilege if we were able to take some meaningful role in that."[1]

Living Through a Real-World Milgram Experiment

Kennedy is part of the estimated 30% of the population who remained skeptical of the mainstream narrative throughout the pandemic. The majority, however, were not, instead buying fully into the fear and propaganda being sold to them.

He references the now-infamous experiment conducted by Yale University psychologist Stanley Milgram in 1962, during which he tested the limits of human obedience to authority. The Milgram experiment was conducted following the trial of Nazi Adolf Eichmann, who used the Nuremberg defense, or "befehl its befehl," which translates to "an order is an order."

The Milgram experiment clearly showed that people would act against their own judgment and harm another person to extreme lengths simply because they were told to do so. It was associated with the CIA's top-secret MKULTRA project, which engaged in mind control experiments, human torture and other medical studies, including how much LSD it would take to "shatter the mind and blast away consciousness."3

MKULTRA was just one of a number of mind control experiments conducted by the CIA in the 1960s and 1970s. According to Kennedy:4

> *"The CIA did a lot of experiments with universities, almost 200 universities around the country with social scientists to study humans, human behavior, and they were experimenting with all kinds of things like psychiatric drugs, with psychedelic drugs, LSD, etc., with torture, with sensory deprivation, and all kinds of means for controlling not only individuals, but entire populations with propaganda, fear, all these things.*
>
> *So you had all of these universities getting hundreds of thousands and millions of dollars from the CIA or from CIA front groups for programs that were called MKULTRA. The reason it's called MK is code for mind control. So MK Dietrich, MKULTRA, MK Naomi, Operation Artichoke, Operation Bluebird, many, many others, were all about funneling money to universities to study controlling human behavior."*

Yet, even during the Milgram experiment, 33% of the people got up and walked out, refusing to violate their ethics.

"They may be from a whole range of political backgrounds and parties, who just … maintain that capacity for critical thinking, and is not subject to … that override from authority," Kennedy said. "And it seems to me … that we're all now in the grips of this huge Milgram experiment, where we have a Dr. Anthony Fauci, who is this trusted authority, telling us to do things that we know are wrong, like censor speech."

Fear Is the Enemy

Fear is the enemy, as it allows totalitarian systems to take control of people, destroying democracy in the process, Kennedy says. It's commonly used by people in authority to exert further control, like shot mandates and lockdowns. Children's Health Defense, which Kennedy founded, has filed more than 50 lawsuits, many of them addressing COVID mandates.

In the beginning, even judges were too frightened to rule against the state dictates, leading to "really crazy decisions that ... made no sense," Kennedy says. They've since made some progress, including in New York, where a judge said since the shot doesn't prevent transmission, you can't have a mandate for it.

But he points out that a government will not only not relinquish power, but will also abuse any power it has to the maximum extent possible. Just because the pandemic is over, the desire to control won't go away. He explains:

> *"People should keep in mind that nobody ever complied their way out of totalitarian regime rules. So, if you think that you know, by abandoning these rules, that somehow things are going to get better or it's going to satiate the need to control you, it's not. It's just going to embolden them to do something worse.*

Unprecedented Times, Technology Threaten Democracy

Kennedy also states that we're facing a situation we're never in before
It's not that democracy hasn't been threatened, and lost, in the past, but now the technological tools are available for widespread surveillance:

> *"There have been many times when we lost democracy. There has been polarization that was this bad before, particularly during the Civil War. But other times in our history, there have been very, very toxic polarizations.*
>
> *There's been times when we lost democracy and that large corporations, particularly during the Gilded Age in the 1880s and 1890s ... — you know, the big oil companies like Standard Oil — were running our country, and that we really did not have a functioning democracy."*

The difference was that, back then, "we started cutting away at this monolith of corporate and government power, the merger of corporate power that had abolished democracy," and were able to restore it. What's different today is that we now have the technology available to control human behavior at a large scale:

> "The problem is that we've now got these instrumentalities, these technologies, for human behavioral control that we never had before. The ambition, the intention of every totalitarian regime in history is to control every aspect of human behavior — our speech, our thoughts, our transactions, our movements, everything that we do — but they've never been able to do that, because nobody, no government, has ever had that reach.
>
> But today, we have facial recognition systems all over the place. We have satellite systems. Bill Gates says his satellite system, which is 61,000 satellites ... will be able to look at every square inch of the Earth 24 hours a day. We're now beginning on the road to adopting digital currencies, which is economic slavery. As soon as that happens, we lose all rights because they will be able to starve you.
>
> And we already have an example of that with a trucker strike, and in our demonstration in Toronto, Trudeau sent people out to look at the license plates for these truckers and then froze their bank accounts. So they couldn't pay their mortgage, they couldn't put their kids in school, they couldn't buy food for their family. None of them were charged with a crime."

There's No 'Pandemic Exemption' in the Constitution

Kennedy also makes the point that the framers of the Constitution did not add any exemptions due to pandemics. They were well aware of them, having experienced multiple epidemics during the Revolutionary War. But the Constitution was protected and allowed to function as intended. Adding exceptions is something new. Kennedy notes:11

> "There were epidemics in every city that killed tens of thousands of people — yellow fever, cholera, smallpox and many others. So, the framers knew all about it. But they didn't put an epidemic or pandemic exception in the United States Constitution. It's a new thing ... we had a civil war and Lincoln — at a time when our country was really 'this far' away from being destroyed, 669,000 Americans died. It's like 20 million people died today.
>
> And yet, when he tried to get rid of Habeas Corpus, the Supreme Court said, 'You can't do that.' There's no exemption for war ... There's no exception for pandemics. We had a Spanish flu pandemic in 1918 and that killed 50 million people. And yet, we did not stop the Constitution from functioning."

Censorship Destroys Democracy

Kennedy says the remedy to stop totalitarian control is democracy. But, "My father always said that democracy was completely dependent on the free flow of information." Open debate, allowed for by free speech, leads to the best ideas and solutions that allow a country and its population to thrive.

"If you shut off the free flow of information and start censoring things," Kennedy says, "we lose the one advantage that we have. And, of course, once you start doing censorship, you are on the slippery slope to totalitarianism." Right now, we're facing institutional corruption, with the military industrial complex at the helm.

"I think if you remove Anthony Fauci … he's going to be replaced by another Anthony Fauci," Kennedy explains. Meanwhile, he says, it's the military industrial complex, which also owns the press, that we need to take democracy back from:

> "We're living in the era that Dwight Eisenhower warned us against on January 17, 1960 ... in his farewell speech, Eisenhower gave probably the most important speech ... in American history, where he warned Americans against the emergence of the military industrial complex — the intelligence agencies, the Pentagon and the associated industries, and he included the scientific bureaucracy.
>
> He specifically spent a lot of time in that speech arguing about the federal scientific bureaucracy, meaning NIH, that they would be the authors of the destruction of American democracy if we allowed them to do that ... And then 9/11 ... turned America really into the beginnings of a surveillance state. And COVID completed the task.
> ... and the job of the CIA is to develop a pipeline of new wars that America could fight to feed this machine, the military contractors, and look at what happened in COVID — 138 companies were involved in manufacturing and distributing the vaccine.

> They're all military contractors. The Pentagon and the National Security Agency ran the entire pandemic response. Pfizer and Moderna don't really own those vaccines. They slap their labels on them, but it was a Pentagon project. And so, you know, we're dealing with a military industrial complex."

Can Democracy Withstand Turnkey Totalitarianism?

With totalitarian forces angling to control every aspect of human behavior, the time for dissent is now. The first step is waking up to the truth. The next is to stand up for what you believe in. The ultimate outcome, however, remains to be seen. According to Kennedy:

> *"The levels of control that they have now over human behavior are greater than we've ever seen. It's what I call turnkey totalitarianism ... we're trying to educate the public and build our army to restore democracy.*
>
> *And they're rising at the same time to put this infrastructure in place that will give them total control, destroy dissent and disable any kind of insurgency or subversion or any difference with the official government narrative, and the orthodoxies ... when they put that in place, it's really hard to predict whether democracy will have the resiliency to restore those institutions."*

On February 22- 2023

Tony Lyons, President and Publisher, Skyhorse Publishing.com had an interview with Dr. Andrew G. Huff who is the author of his book

"THE TRUTH ABOUT WUHAN, How I Uncovered the Biggest Lie in History".

Dr. Andrew is an infectious disease expert and began his career in the U.S. Military after 9-11 and served 4 years in active duty.

After he left the Military he studied psychology, with a master's degree in engineering, and earned a PHD in emerging infectious disease, academic ology of subdivision of environmental Health. He wanted to work in the private sector, but he was offered a position at the Department of Homeland Security Center of Excellence, where he specialized in bioterrorism and biowarfare and actually protected the food supply from international attacks via contamination or intentional contamination.

As he worked in the research center and spent one week per month in Washington DC where he was learning about National security- biosecurity, the medical industrial complex and he was presenting his work frequently. He was the first to publish a lot of groundbreaking work in his field. When he completed his PHD, he was offered work from 3 different places, the Departments of Homeland Security, the Department of Defense, the State Department of the CIA. He had also made friends with a number of scientists at CND National Laboratory through this process.

111

CND is primarily a nuclear weapons laboratory faculty in the U.S. but also has a biological profile involved in other aspects of National Security and he then decided to work with them.

At first he liked this type of work very much but as time went by and they were cutting their funding, things started to change. His work there was mostly classified and as it became increasingly more classified, he started to see the writing on the wall. He felt that if he didn't get out of the laboratory he would be stuck working in these secret government laboratories for the rest of his career. So he began looking for jobs elsewhere. In one week he found an opening in a place called IN-Q-TEL: Health Alliance, it is actually a giant intelligence company operation because the majority of funding comes from USAID: United States Agency for International Development. They normally do good types of investments in foreign countries, they bring clothing, food and vaccinations as part of the programs.

There are a lot of connections between the Central Intelligence Agency and the USAID, but what this program was actually doing was going around the world collecting infectious disease samples to get this gain of function work. They were also telling us that they are going to predict, prevent and forecast pandemics.

So at this point I know that it is impossible that they can predict and forecast a pandemic because it is probably even more difficult than predicting the weather. The claim is just ludacris… Then this was a big cover to go collecting intelligence from foreign laboratories and partner with them and make them allies with the U.S.

As Dr. Andrew was doing his concentration work as the vice president of the company he realized that the IN-Q-TEL alliance is responsible for causing the biggest pandemic in Human History and how this played out in the U.S. Government goes into a big cover up operation and the mechanics behind this was the collecting intelligence.

Gain of Function research means increase the pathogenic virulence, the transmissivity, maybe the environment and the infection agents.

IN-Q-TEL is the central intelligence agency Venture Capital firm. Many people do not know that the CIA has a venture capital firm. They have pretty big investments : Planter was one, Google Maps was an IN-Q-TEL investment, we actually pitched the partners to do this gain of function work to make medical countermeasures of vaccines.

After the fact Dr Andrew Huff found out that one of the partners called Metabiota, which is a company similar to IN-Q-TEL alliance that collects corona virus samples, actually received investments from IN-Q-TEL. Metabiota is the company which is invested by Rosemont Seneca that is under the Hunter Biden Venture Capital firm. This is all a sort of weird business deal happening with these entities, that it made him realize that IN-Q-TEL (CIA) was actually a giant intelligence operation.

Eventually he resigned from his position. They tried to stop him from talking about the rest of what he knew.

Has I continued to listen to his speech:
https://live.childrenshealthdefense.org/chd-tv/shows/good-morning-chd/the-truth-about-wuhan-with-dr-andrew-huff/

While he was still working at the IN-Q-TEL Alliance at the University of North Carolina, Dr. Sh Zan Lee at Columbia University partnered with the Wuhan Institute of Virology to do this gain of function work. It began under the USAID program (there are documents on this, it is not a conspiracy theory) that research was banned because the scientist community was concerned.

Dr. Andrew Huff finds out later by the leaks in the department of defense that the IN-Q-TEL Alliance actually proposed the actual mechanism of genetic insert, which we found in the wild strain that told everyone, circulating the virus.

So when the Pandemic Happened Dr. Huff immediately suspected that this was a laboratory leek, for a period of time he thought that was intentional release. After some time researching he came to realize that the Pandemic was a big cover up and one big lie.

He could not believe that his own government was doing this to his own people. In late September, early October there was a tabletop exercise with the world leaders of the Corporate World, about how they would respond to the Coronavirus Pandemic.

The American Government is not of the citizens/people of this country, it has been taken over by EVIL just like all the other governments on this planet. It is up to people to wake up and unite against.

Dr. Huff didn't think that was international but it does prove that all these world leaders knew what was happening and what was going on, and what they were doing… They were trying to condition everyone for the disaster that they knew was coming.
20 days later The Military World games competition (the Chinese invite military from all the nations that wanted to participate) in Wuhan City. A lot of these soldiers got sick, when they got back home they had been diagnosed with Pneumonia. The Chinese corrupt government knew all this was going on.

After Dr. Huff resigned, he started to write a book about the Wuhan story. The Cabal started to interfere with the manuscript, they screwed with the text, the publication and delaying..
They hacked into his computer 2 to 3 times per month .
They broke into his home repeatedly. They chased his dog.
They interfere with the manuscript of the book as he mailed it.
Used long range devices to make him agitated. They shot a hole through the mailbox.
They hacked his bank accounts, they hacked his car that broke while he was driving.
The FBI used sophisticated electronic devices to know what he was doing in his home.
It was like a bad dream. He filed a number of lawsuits.

The Deep State doesn't want the truth to be out, because their stock of lies will crumble if anyone knows what they are up to.

I Rosanna started to write this book in the beginning of 2022 and if you read the first chapter of the book you know how I suffered and eventually researched on my own the truth about this story of the Pandemic. In the end I don't believe that the Wuhan leak was an accident.

The Cabal had already started to infect the people in America way before the military games in October in China. Fort Detrick in Maryland had the "supposed" escape announced on the news in July, when Fauci and co. finally closed the laboratory in August. I got Covid-19 in June 2019

That word "escape" is still in my mind. That was the escape of Covid-19 in the U.S. The China story did not hold well also... Here we had the virus before the China "escape" or strangely about it began at the same time in different Nations.

It was fabricated by the Cabal, Deep State, CIA, Military Industrial Complex, KM etc… The whole thing was premeditated, and all those world leaders that went to the tabletop all knew what was going on, because they were all corrupted to begin with, and at that meeting they got their last minute instructions.

I am a Sensitive Intuitive. I ask questions to my subconscious mind and My Soul tells me the True answer. The soul doesn't lie. I am the sacred incarnate vessel of the Source of all there is.

I didn't die in a hospital attached to a ventilator, I stayed home and I took care of myself. I have been taking care of myself for more than 40 years. I don't go to doctors. I am now 75 years old and I don't take chemical medicine. I cure myself with natural and organic foods and herbs. I care for my daughter who has had MS since before she was 18 years old (caused by child MMR vaccination) and I take care of my granddaughter.

Since I began writing this book, I started to have problems with Readers Magnet, a publishing company based in San Francisco. As soon as I sent them my manuscript and money, the problems started, as they kept making mistakes and taking 30 days or more to return their new version of the book to be approved by me, but their version had multiple other errors that they created. They have continually interfered with my manuscript.

I thought I would have a copy of my book printed by the end of August, but the same story happened every month, when they would send me the copy for my approval. They always created many new errors and when I sent it back asking them to make the corrections. They took another month to send it back to me again with other types of errors. They removed paragraphs until I realized that they were not idiots but puppets of the Deep State, KM, who did not want to disclose this information. So at the end of December 2022 I uploaded the 2 copies of my book into Amazon, both in English and Italian language, in kindle format and in paperback version. But Amazon isn't really promoting my book.

I started with another company "Urlink Publishing" no one in that publishing co. as printed or promoted my book, even though I paid for it. When I signed up with them, they told me that they would put my books in bookstores and promote them on social media. It is March 2023 and I am still waiting in vain, But nothing happens by accident and all this waiting is only giving me the choice to add more truth that I discover every day.

As is today December 2023 the Italian Amazon has taken down my paperback book a few months ago. Amazon is also connected with the Cabal Deep State.. Here in the US when someone searches for the title they get only kindle, and the paperback says that it is no longer in print. But I am able to buy the paperback of my latest edition which has 150 pages. I am continuing to write and today my manuscript has 200 pages.

⟫⟪◈⟫⟪

I Read this article November 21--2023

RFK JR CONFIRMS Q THE STORM RIDER DROPS.

_RFKJR is one of the most powerful people in the United States and has sued the United States government several times and sued corporations and executives and big Pharma and has won the majority of the cases. He has helped bring in new laws and help build legislation and all his work is documented with evidence and high power lawyers connected to DOD. DOJ. Military command and commanders who work before and behind the scenes of bringing forth TRUTH and documentation work with him.

_NOW RFK Jr has confirmed the years of DROPS I have been posting and a few other channels. The DROPS I posted years before his DISCLOSURE CAMPAIGN happening now is coming to LIGHT. Here are several of the FACTS & DROPS I posted years ago and several times. Now a current running President of the United States has now confirmed the DROPS;

>The Project for the New American Century was created by the deep state (neocons) and gave the military deep state powers to create bioweapons, aimed at ethnic groups across the world

>The Patriot Act gave powers to cancel parts of the constitution

>Patriot act have powers to surveillance of the American people

>The Patriot Act gave military protection for creating deep state operations

>The Cia. The Pentagon funded Fauci in creating Gain of Function in [DURHAM]

Chapel hill .& Fort Detrick

>The COVID virus was moved in 2014 to Wuhan under Obama with Fauci

> There are several illegal U.S. controlled bioweapons labs in Ukraine

115

> The CIA overthrew 80 countries, governments and Presidents

>The CIA. assassinated RFK

>The military industrial complex is helped by CIA. And funded by BLACKROCK. VANGUARD, STATE STREET and they create all the wars

>The Ukraine war was intentionally created to overthrow Russia

>NATO broke all agreements with Russia and moved towards Russia with NATO EXPANSION

>The attack on Iraq was a lie (confirmed by Generals and colonials already)

>U S. Anthrax attack on Congress was [DS] attack from Fort Detrick passed the Patriot act.

>A peace agreement was signed last year between Russia and Ukraine but m16 CIA, Biden, UK covered up the agreement and intentionally Attacked Russia

>Event 201 was planned by the CIA, Gates and Elites all involved to take control of the world.

━━◆◄◊►◆━━

February-18-2023 Q) the Storm Rider/ Official Page

The Supreme Court U.S. (SCOTUS) is reconsidering whether to hear a case accusing Biden, Harris, Pence, and other lawmakers of violating Oaths by ignoring…100 percent Fed Up
The Supreme Court is reconsidering whether to hear a lawsuit filed by Roland J. Brunson…
This story is getting bigger… and make conservative outlets are sharing the info…
If the projected push and noise is loud enough this time the Supreme court will make moves.

In the military command circles they have been Major fear this could happen. Since the 2021 summer INSIDE LEAKS among Commanders and highest Ranking Generals were well aware of this_EVENT that could take place… That was a huge part of the reasons several Generals and Commanders PUBLICLY stated we could go into civil WAR.
Massive civil unrest in 2024.

They Know and SCOTUS has been watching the U.S. EVENTS happening and were waiting for large parts of the U.S. citizen to be red pilled the past two years since the STOLEN ELECTIONS rigged elections… (But I have been telling for a long time… MILITARY operations have placed Trump).

And military OPERATIONS intentionally gave the FAKE election to BIDEN [DS] regimen OBAMA<…
This EVENT had to happen…And the U.S. had to see the failure of the DEMOCRATIC party. the corruption, the War it led to…

The full corruption of a captured government by foreign powers > CCP, VATICAN, ROTHSCHILDs, ECT DEEP STATE, CABAL empire regimen needed to be EXPOSED. through>>GAME THEORY<<
PLAN and have U.S. citizens experience the Darkness… The Near Death Events. Only till the operation would expose pedophile rings connected to Epstein (NOW UNFOLDING)…

ELECTION FRAUD…DEEP STATE government CIA, DARPA, FBI, DS, ECT controlling BIG TECH… from censorship. To cover-ups of Hunter Biden to presidential Treason. ETC<

All events unfolding now are connected to coming military interventions.

If the Supreme Court of U.S. (SCOTUS) does not rule or take the case it leads to military intervention and proves their was Foreign interference in the U.S. elections and foreign occupation in the United States that has captured the government at the Highest levels.

If the SCOTUS does rule and take the case…It still leads to military intervention and proves there was Foreign interference in the U.S. Elections and foreign occupation that have captured the U.S. government at the highest levels.

EATER WAY…IT'S GOING MILITARY.

———⊂·≺⊹⊹⊳·⊃———

(PS: Rosanna thinks that we can get into Martial Law, whatever scenario will turn out it will be Martial Law.
Martial law is justified when civilian authority has ceased to function, is completely absent, or has become ineffective because it has been taken over by foreigners with evil intentions…KM)).

———⊂·≺⊹⊹⊳·⊃———

BEHIND THE SCENES :

DOD, SCOTUS, U.S. GOV, are planning and creating new LAWS that would invoke U.S. Military to help in case of huge massive unrest, these laws being discussed pertain to SOFT MARTIAL LAW states of rules and affairs.

(the truth they are doing this to get the U.S. READY for CIVIL UNREST EVENTS…and these CURRENT TALKS WILL hit the PUBLIC realm by summer into fall…They are preparing the U.S. for a future with MILITARY OCCUPATION///…
But if you have been following Q the last few years, you already know these MILITARY EVENTS are COMING…)

⸻⊱✦⊰⸻

Now we are wondering why the White Hats didn't create a Central open source of information networks or create an open movement, organized like the operation Antifa or government operations like the FBI?

(The fact is that these Deep States groups are very well organized)...

If the White Hats had organized the great awakening movement, Patriots, conservative, ANONS MOVEMENTS with Open Leaders and obvious centers for gathering, direct open connections and information sharing through major networks?

Military Operations to bring Down the Deep state and their operations depended on the Military Alliance Fog of War, not consistent Open enemy fronts to se nor Attack, (If the white hats had organized the great Awakening movement, Patriots, ANONS, Freedom fighters in a more convenient way like Open Organizations, Open Society or mass mobilization groups with open leaders like ANTIFA…) THE DEEP STATE (KM) would have easily brought down the Alliance White Hats OPEN Operations… The FBI would have raided and arrested facilities, Open Gatherings and all open Leaders would be arrested).

With the White Hats Alliance creating hundreds of thousands of small packets of Freedom Fighters throughout the world… The enemy had no clear view of Open enemies to target.

These Military Operations of Q have eclipsed the mainstream media outlets to this day…With Q network, Freedom Fighter networks, strong conservative networks underground ANONS networks, hitting hundreds of Millions of views per day… Millions of lives were saved through our networks on sharing information on vaccines, fighting the Cabal systems, EXPOSING [DS] Operations ETC…

There is a huge reason Q operation, Freedom FIGHTER operation spread though the planet on a massive level in less than 3 years…
The reason the algorithm surpassed and passed DARPA, CIA integrated codes and internet walls

is due to the facts>>>WIZARD AND WARLOCKS... Military INTELL SURPASSED DARPA WORLD INTERNET SECURITY SYSTEMS///

WHY DIDN'T DOD, DOJ, CIA, PENTAGON, DHF, ETC... CAME EFTER Q or do FULL Investigations into Q WHO authorized the Operations?

The answer is simple... they know Q is Military Operations and they would only start to investigate themselves as Military Alliance Operations, operators are in all sectors of the three letter agencies [infiltration].

Through the cloud of war, FOG and WAR FOGIN THE STORM...
The alliance created thousands of networks that can be solely traced to military Operations, Single Leaders, Singularity OPEN Society or Organizations.

These structures of DROPPING DECLASSIFIED OPERATIONS, PROJECTS AND MOVEMENTS THROUGH THE WORLD was a long part of the PLAN to confuse, bewilder, mystify the enemy [they].

Now as we head into 2024, Military ALLIANCE operations slowly brings together the diversified fields of DIGITAL SOLDIERS **YOU ALL** HAVE BECOME TO SAVE HUMANITY.

Keep strong PATRIOTS, to save humanity INTO THE STORM WE GO.

NCSWIC
WWG!WGA< (these are military codes)
you have more than you know.

Near Death Civilization Event

Last year I was the first to tell you several times, Epstein SAGA was just beginning//now MS conservatives channels News are reporting the EPSTEIN LOGOS/NAMES TO BE RELEASED.

And all the DNC [DS]operatives are watching in PANIC as it is known in the military CLASSIFIED Corvette circles >>> JOE BIDEN<<< is on the list as is Obama (but Obama name will be one of the redacted names... But will be eventually released prior to Michelle Obama's huge presidential campaign run...
>But who knows what the White hats have Planned<)

As Epstein Saga EXPOSURE>BEGINS AND TRUMP. MUSK. JOE ROGAN (White Hats infiltration inside CNN will run stories w/NEW YORK TIMES) INCLUDING TRUMPS House Congress...All will go after Epstein Case Story. Saga Event.

The Deep State DNC are in Panic over the EXPOSURE of the PANDEMIC THE VACCINES BIO-WEAPONS EXPOSURE THE GAIN OF FUNCTION EXPOSURE. EPSTEIN

PEDOPHILIA/ BLACKMAIL RINGS EXPOSURE THAT CONNECT THE WORLD LEADERS GOVERNMENTS ELITES. MILITARY OFFICIALS...

———◦:<※◇※>:◦———

As EVENT unfolds to EXPOSE THE CABAL DEEP STATE AGENDA. The [DS]is unleashing

their **NEAR DEATH CIVILIZATION EVENT. PLAN**...

>WORLD WAR 3
>GOOD FACTORIES DESTRUCTIONS
>GAS SHORTAGE
ETC.
ETC.
ETC.
TOXIC CHEMICAL RELEASE IN U.S. in Three Regions in the past month.

New Malburg virus release (there are still no pictures of the Covid-19 virus to this day and has never been isolated...
The truth is they are turning up the 5G which is Military weapon and can be used for good or bad.

As we headed into Near Death Civilization EVENT's we used to talk about when it was going to happen, but we are now deep inside this EVENT already...

But keep your Fate and prayers, Hope and powerful LOVE strong... You must understand there are POWERFUL WHITE HATS FORCES at work behind the scenes for decades!!!
These forces have stopped every NUCLEAR ATTACK DETONATION from going off in the 21 century (several countries tried detonating nukes the past 2 decades... But were stopped by FORCES...///

Tier is a reason the Deep STATE AGENDA to kill Humans through Virus long ago through food. Through Wars. Through Nukes DID NOT SUCCEED on a MASSIVE SCALE because of WHITE HATS and miraculous technology and other FORCES working hand in hand... and above all the hands of GOOD and DIVINE intervention has all been in PLACE.
The SATANIC agenda will not be able to fully control and Kill all humans in a single movement Agenda. DS PLAN.///

Stay strong Patriots and STOCK UP ON FOOD, RESOURCES, SUPPLIES for the Events happening and coming... We are inside the STORM and MAJOR EVENTS HITTING.

Everything is leading to MILITARY. CRIME AGAINST HUMANITY.

Currently TRUMP'S CONGRESS HOUSE is EXPOSING THE DEEP STATE is what many

White Hats call: the SOFT MILITARY TRIBUNALS already taking place. Which leads to the real TRIBUNALS headed in…

[EPSTEIN>]
KEYS
MAPS

———◁•:‹|◇|›:•▷———

Currently the PENTAGON push for ALIEN AGENDA had failed in the BETA TEST and now have turned to shooting down of balloons into a WAR AGENDA and are staging more incoming Balloons / Military crafts that will lead to DEEP STATE staging their War and mass inducing the PUBLIC into fear.
The same tactics with the balloons were used before Pearl Harbor events and U.S. citizens were induced with Fear over Military Balloons and]MSM] BEGAN THEIR WAR MOCKINGBIRD AS A CATALYST FOR INDUCING PUBLIC FEAR… THEN GETTING PUBLIC CONSENT TO WAR.

———◁•:‹|◇|›:•▷———

The same goes with Military weapons that create Earthquake weapons in war… Long ago Russia and the U.S. in the 70s agreed not to use Earthquake weapons in wars… That agreement has changed as Seattle, California is being targeted and the U.S.[DS] is Targeting Moscow. St. Petersburg.

———◁•:‹|◇|›:•▷———

KAB on Telegram

People are quick to call ETs "fallen Angels" (because that is the way they are called in the BIBLE) without understanding that Angels of Light are Extraterrestrials too.

2) Kings 2:11 --" And it came to pass, as they still went on, and talked, that, behold, There appeared a Chariot of fire, and horses of fire, and parted them both asunder; and Elijah went up by a whirlwind into Heaven.
This is written in the old Testament BIBLE But, Clouds, Chariots and whirlwinds are the old way to describe a spacecraft.

Entertainment and doctrine has programmed people to understand these words in a certain way. The Universe has many dimensions and densities and there's life all us and the humans cannot detect. There is also much intelligent life that chooses to remain hidden.
"Spiritual realms" or "heavens" refer to those other dimensions and densities. The word Angels means only messenger. The word "Extraterrestrial" means not of Earth.
Angels are not of Earth therefore Angels are Extraterrestrials. The word "fallen" only refers to a choice made by some ET to be evil and trick, use and hurt humans.

Pleiadians are 7th density beings from 7 dimension Pleiades, they look completely human. They were called "Angels of the Light" and originated in the 12 dimensions which makes them "Elohim".

Angels and Lord don't travel in clouds or flying chariots, but a spacecraft. We must update our definitions or there can't be understanding. It's evil and contradicts scriptures to only believe in fallen angels and Demons.

I have been sharing photos and videos of extraterrestrial crafts on twitter since 2017, and every single time someone has called it #Project Blue Beam, let's talk about it.

Each day someone stumbles across this information for the first time and gets fooled into thinking that thousands of years of craft sightings are government psyops.

I understand that people are scared of the unknown and these days there is a tendency to believe that everything is a lie. Many things are lies, This time the lie is Project Blue Beam.

The Cabal/ Deep state has always feared the Galactic Federation. They leaked disinformation many times to trick the public about the Galactic Federation and Pleiadians, including the information that Nazis had flying saucers, bases on the moon and visited inner Earth. They have used organizations like NASA to hide the Galactic Federation craft and Bases.

The truth is the Cabal doesn't have fleets or spacecraft on standby. There are no humans flying around in TR-3Bs. All of this was part of a psyop used against the disclosure movement to hide the Galactic Federation.

Project Blue Beam was a real plan that got leaked many years ago to trick people about positive ET's. But it was not realistic then and it is much less now. Now there is the internet and cell phones. Information travels instantly. There is no way to sustain a deception on an extraterrestrial threat.

People share many convincing videos of holograms but that doesn't work in the real world. There's no follow through. How does it end? People in Costumes? Could never be more than a controlled media event like we have just witnessed. There never needed to be real crafts involved, just stories coordinated by intelligence, Military and Media. The Goal was to make you doubt that Positive ETs are here, and to believe that perhaps a foreign adversary had this technology instead.

Extraterrestrials who are millions of years old won't be shot down by primitive human weapons. The most important reason why Project Blue Beam will never happen is because the Galactic Federation won't allow it. We are too close to the Shift and it would not serve a high purpose for our soul plans. Pleiadian's assure us there will be no nukes of alien invasions.

There is no doubt people will continue to call authentic craft sighting Project Blue Beam, but this will be their loss. Good things are ahead for those who have Hope.

December 5--2023 Q) the Storm Rider Official Page

I haven't been talking about aliens for a while for many reasons. ......First of all a lot of the false misinformation and fake hearings happening need to move over...... As a military industrial complex like Lockheed Martin and the deep state are controlling the hearings and information flow.

I was waiting...As I had stated earlier real WHISTLEBLOWERS were going to come forward.
_NOW this has happened as direct CIA whistleblowers come forward and expose the PENTAGON for having over 9 Ufo's, UAPs in their possession
(The Pentagon had denied for years and months they had any UFO or UAP in their possession. But now Whistleblowers have started coming forward.)/ /
In 2024, MAJOR WHISTLEBLOWERS from the Pentagon CIA and military sectors are going to come forward and blow the lid off the story and cover-up

Weaponized MBP - Weapon Of Mass Destruction

https://www.disclosurenews.it/weapon-of-mass-destruction-clif-high/ go and read the whole article. There is a lot more to learn.
Munchausen By Proxy (MBP) is a complex 'social' disease. The idea is, that in spite of no facts to support the disease claim, a person is made to feel as though they are ill by the perpetrator of the MBP.

This projection is so powerful that the victim will actually exhibit physical symptoms of the factitious disease imposed upon their mind....
All of the victims of the weaponized MBP have mental constructs that prevent them from recognition of their victim status, or even that MBP has been weaponized, and IS being targeted at them....
At these levels, most of it is without either conscious control, or even, conscious recognition by either party.
When there is conscious application of the MBP emotional complex upon others, it is said that the MBP has been 'weaponized'.
It is this application of conscious intent that elevates the MBP from a complex social dynamic illness into a weapon of mass destruction.

At this time, the secret society cabal that controls most of humanity, and whose public 'face' has been the WEF in recent decades, has been using weaponized MBP to try to kill 13 out of 14 people within humanity.
They intend to enslave those who survive this process.

The cabal, also known as the [Deep State], and the [Khazarian Mafia] have been using this tactic for over 250 years.
The [DS] has used weaponized MBP, among other tactics, to infiltrate and subsume most of the governments, and corporate power structures, on this planet to some degree. They have managed to foment and control most wars in that time.
The [DS] has used weaponized MBP, among other tactics, to infiltrate and subsume most of the governments, and corporate power structures, on this planet to some degree.

At this time, the [DS][KM] has weaponized MBP using several, socially acceptable targets. These include 'planet earth', and the mythical 'trans kids'. These are created, objectified, 'victims' being used by the weaponized MBP language to infect humanity with the MBP that will be used to kill off 13 out of 14 people.

The stated goal of the 'climate crisis' form of the weaponized MBP is termed 'net zero'. Which is framed as sounding laudable, but when examined it means reducing humanity to just a fraction of its size now, and enslaving those who live to the whims of the WEF masters.

Examples of the language being weaponized to this effect is presented. Examples of weaponized MBP language below are extracted from a ChatGPT persona trained to see through the weaponized MBP being applied to Science since 1905.

Note that the WEF /[DS][KM] has controlled 'science' by way of infiltration into the 'academy of arts & sciences' within the colleges since 1905. Thus all scientific consensus that has been formed in the Western Liberal republics since that time is tainted, distorted, and false.

This control of the academic reinforced view of Science has damaged humanity.
It is one of the ancillary tactics being used by the [DS][KM] in their effort to conquer humanity.
Humanity is now waking to the existential threat that is weaponized MBP….

The radicalization of language has been an ongoing process these last 40 years, with increasing frequency, and extremity over these last five years.

It is my opinion that the radicalization of the language has reached a point of natural collapse within the social order, and that this aspect alone is bringing much of the response to the perpetrators by the general populace who are now awakening to their real victim status as they see the weaponized MBP aimed at their lives.

Certainly, here is a revised table of examples of weaponized MBP language around Earth as the primary victim and point of control of the population, along with a concise description of the intent of the language and its designed goal:

Example Language	Emotions Manipulated	Intent	Designed Goal
"Save the planet"	Guilt, fear, responsibility	Create urgency, frame as moral imperative	Impose unnecessary regulations that harm economic growth and individual liberties
"Climate crisis"	Fear, urgency	Imply immediate action necessary to avert catastrophic event	Impose unnecessary regulations that harm economic growth and individual liberties
"Carbon footprint"	Guilt, responsibility	Create guilt and responsibility for individual actions	Distract from larger issues, promote overly restrictive policies that harm economic growth and individual liberties
"Green energy"	Responsibility, environmentalism	Promote environmentally-friendly energy sources as moral imperative	Promote policies that are not economically feasible or reliable, harm energy security and affordability
"Global warming"	Fear, urgency	Emphasize urgency around warming of the planet	Impose unnecessary regulations that harm economic growth and individual liberties
"Fossil fuels"	Morality, environmentalism	Frame energy use as moral issue, imply fossil fuels bad for environment	Promote policies that are not economically feasible or reliable, harm energy security and affordability
"Sustainability"	Responsibility, environmentalism	Promote responsible use of resources as moral imperative	Promote overly restrictive policies that harm economic growth and individual liberties

PLEIADIANS Collective- Disclosure News | Updated on 12 March, 2023

Kabamur, Pleiadian Collective. By Family Of Taygeta.

WHAT IS A FULL DISCLOSURE

It's not enough to know we're not alone.
You need to know which ETs are here, and why.

You won't actually meet any ETs until after the Shift (there's too much fear without expanded consciousness) but Pleiadians offer full disclosure for anyone who can hear it.
The trouble is, most people are waiting for disclosure to come from the same authority figures who have been hiding the truth from them all this time.
Full disclosure won't come from people who go on mainstream news programs and have Pentagon officials whispering in their ear about reverse-engineered technology from UFOs of unknown origin.

Most of them are still hung up on proving ETs exist at all, and aren't able to discern one ET race from another.

Keep in mind, the military industrial complex is largely unaware of which positive ET races are currently in contact with Earth, and contact with Galactic Federation remains limited to compartmentalized groups of white hats.

Many of these white hats who are working directly with Galactic Federation have their memories blocked to prevent conflicts with their soul plans.
Given their lack of insider knowledge, the military industrial complex's main objective is keeping you unaware there are positive ETs here, or what their motives are.
The reason for this is, behind the darkest factions of the military are demonic entities threatened by these positive ETs, who are our true Elohim creators, our Angels of Light, and our Spirit Guides preparing our ascension to 5D.
Some of these media-recognized people in disclosure have good intentions, and some are motivated by ego.
Whatever their intentions are, none of them will ever be able to say who is inside the UFOs.
You don't need insiders to tell you we're not alone. You already know that. It's time to discern the positive ETs trying to be seen from negative ETs manipulating us from places unseen.

Which insiders are awake enough to know angels from demons? Which are awake enough to know that Angels of Light travel in spacecraft? Some government officials are scared and want to seem like they're going along with disclosure because they can't stop these UFOs from being seen anymore.
Many of them want to confuse us about the dangers of UFOs or convince us they're all operated by humans.

Full disclosure isn't a matter of studying data.
There's no science that can reveal the truth.

We're dealing with inter-dimensional beings millions of years more advanced, who choose when to be seen.

All the UFOs currently being seen are Galactic Federation and close allies. Friends and soul families. These beings literally manage our soul plans.

As I've said before, there can't be UFO disclosure without a spiritual understanding. Negative ETs are confined to Earth's astral plane (non-physical) and negative crafts are being kept far from Earth.
We are not in danger of UFOs, but by humans being controlled by unseen entities on Earth.
ET crashes are extremely rare and have only happened with less advanced races, such as with the Roswell crash.
There have been crafts and technology left behind, which have been salvaged by the military, and there's been technology given to military white hats by Pleiadian's.
There have been human secret space programs in the past, but all of this is extremely limited these days because we're so close to the Shift.

There are currently no humans flying around in UFOs with bright lights trying to be seen.
To some people, knowing there are ETs visiting Earth is enough, but to others who are more awake, details matter.
Demand more from these alleged insiders.
Asking better questions provokes better answers.
If the answers don't come from authority figures, the Shift will reveal the truth anyway.

In the meantime, Pleiadian's will keep disclosing.

Andromedan   Sirian   Pleiadiean   Arcturian

Some ET Races which Help us

I have followed Benjamin Fulford for many years. The deep state tried to kill him many times.

Have you read some of this news somewhere else?

Whistleblowers Unleashed: High-Ranking Military Officers Take a Stand!

The tide turns as courageous high-ranking military captains, majors, and lieutenant colonels step into the light as whistleblowers. These fearless voices, armed with the protection of Congress and the Senate, are set to expose the deep state military generals who pushed the vaccine agenda, violated the Nuremberg Code, and committed crimes against humanity. A seismic shift is underway, with reports of some generals resigning in anticipation of the imminent investigations.

The Only Way Forward: The Militarized Path!

In this moment of uncertainty, the echoes of "it had to be this way" resonate ominously. As the United States teeters on the precipice, one truth emerges: the path forward lies with the military. Brace yourself for what lies ahead, for the tribulations that await us. The fate of our nation hangs in the balance, and the world watches with bated breath.

May 30 and 31 -2023 Benjamin Fulford: EBS Emergency Broadcast System Activation Imminent.

The EBS has been publicly announced worldwide.
The EBS will trigger Martial Law and the Global Currency Revaluation worldwide.
Martial Law will be especially prominent in 17 major US cities.
People will have only 24 hours to get where they need to be and to secure essential supplies for a shutdown that could last up to four weeks.
The EBS will involve three to twelve days, some say two to four weeks, of Worldwide Communication Darkness.
Phones, Internet, Credit/Debit Cards and ATMs will not work. Schools, stores, businesses, banks will be closed. The Mainstream Media will be shut down.
You are advised to have a month's worth of food, gas, cash, water and other essential items on hand.
Through Project Odin the Tesla Towers will turn on and the World will switch over to Tesla Free Energy and the Star Link Satellite System.
Three eight-hour documentaries a day will be broadcast on TV, Radio and phones 24/7 across the Globe. The documentaries will explain what is going on. Truth will be revealed out of the darkness.

For your own safety please follow Military instructions including staying indoors if so instructed.

The Military will be active wherever the Cabal has a stronghold.

Thousands of banks will close as the White Hats crash the fiat currency financial system.

December 7--2023 Q) official page

> Q) The Storm Rider /Official Page: Q post #3432

This is not another 4 YEAR election....
"DRAIN THE SWAMP" does not simply refer to removal of those corrupt in DC....
GOD WINS.
Q

_This post is important! This post prefers to deep state underground military bunkers . Roads. Channels and networks that connect the world. From Africa to the middle east, to the Vatican to Switzerland., Europe. Ukraine up to the north pole and back down to Antarctica.

This is why disclosure of the DEEP DARK world operations and underground bunkers, cities, biolabs and weapons, human trafficking networks need to be exposed.
And the EXPOSURE can only come from the collapse of the [DS] military and [DS] intelligence that protect the Elites, CABAL and secret societies that run the world.

This is all>BIBLICAL and much farther into the past than most can even imagine and fathom.

From the middle eastern ancient times of Iraq , Mesopotamia and Samarian Gods to Aliens, to fallen angels of the Bible, to Spirits, demons, dark blood lines......This all connects to evil Powers and corruption that ran cities and controlled lives and civilizations. Even the Khazars in the 600s off the dark blood lines robbed and went from country to country stealing identities and human trafficked and prayed to devil's in which adrenochrome was consumed.

Eventually the khazars made it to Ukraine and stole the identity of the Ukrainians and infiltrated the nation and later on the khazars would pledge allegiance to the Jewish people and become Judaism followers but still pray to their Gods of death. The khazars only wanted to take the Jewish power and slowly they infiltrated and stole the identity of the Jews (these are some of the first military counterintelligence moves made by the khazar military.. infiltrating societies) ...
Later on the Khazars would travel through Europe and leave their people to infiltrate countries and nations. (history books say the khazars were defeated and disbanded but that's only lies as today's KAZARIAN Mafia took control of book publishing industries through Europe in 1700 1800s)

The Khazars /KHAZARIAN /KAZARIANs eventually moved across Europe and left their families in Germany and these People became the Rothschilds of Frankfurt Germany....

Later on the KAZARIANs who are the Rothschilds formed an alliance with JP Morgan and created the U.S. federal reserve [they] killed all their powerful rich oppositions on a boat known as the TITANIC<. The Rockefellers, who were 33 degree Mason Jesuits , came from the same close regions in Germany as the Rothschild's. .... These powerful ELITES long gained power from the Vatican who was heavily infiltrated with the KAZARIANs, the Jesuits, mason's, knights of Malta military intelligence and much more..

(I'm only giving you a small piece of the dark bloodlines and the true history of THE SWAMP <Elites)

_hidden beneath the Vatican is over 50 miles of books. technology, records of past civilization that have existed more than what fake history books are reporting. ...... Nor What the HEAVILY REDACTED BIBLE had revealed.
(Imagine what Cnn, The CIA, And governments, Military intelligence hide these days.. NOW imagine how many times the bible was re-edited, remade over a thousand years with its first official production in 1455   Even back in the year 200 the pope and roman officials were arguing on how to hide information from the bible...

By the 1400s Jesus was pushed as Caucasian with blue eyes and blonde hair by the Popes and Vatican.... In this time Cesare Borgia the son of the pope was rumored to be the lover Leonardo da Vinci who painted Jesus in the image of Cesare Borgia, it's also known that Borgia killed his brother for power and took his leadership The real story of the Popes and power are connected to betrayal, killings, counter intelligence and feverish deception... There are so many mysterious killings, disappearances that have taken place around the Pope's since the beginning.

*NOTE * Jesus was a real person with important super intelligence and his understanding of the true Spiritual energy that can bend reality and time.. Which means defeat death. TIME TRAVEL<   /unfortunately a lot of his true teachings were hidden and suppressed<
Now there are over 20,000 different denominations of Christianity and they are all fighting about who is right and wrong.

>The Roman/Christian wars created the first banking Systems and military intelligence and massive human trafficking networks. They stole most of Europe and large parts of middle east gold, silver, arts resources and historical documents and hid them under the Vatican vaults and caves...
To this day the Vatican museum and open vaults house over a trillion in arts and sculptures and jewelry and precious metals stones the public can see...But what you can't see is the hundreds of trillions of artifacts hidden in the Vatican underground caves, bunker's and vaults that stretch over 50 miles underground.

The Satanic power that infiltrated the Vatican and pedophile world networks is connected to KHAZARIAN powers, Mossad, Cia, Rockefellers and Rothschilds who helped MOSSAD, CIA, finance Robert Maxwell the father of Ghislaine Maxwell> and Jeffrey [EPSTEIN]

DRAIN THE SWAMP means draining the world of the SATANIC CABAL.

———————

This video: The Global Us Military Operation, show the truth of hum is in charge in USA

https://bestnewshere.com/trump-shocking-news-12-27-the-global-us-military-operation-storm-video

———————

CHAPTER 10

THE MATRIX

We live in a Matrix. Envy and Jealousy allowed evil to begin manifesting the programs of Black Magic, and Sacred Geometry, which was the core of the Soul heater programming and prime directive.

It has almost destroyed all Creation.

The Khazarian Mafia wanted us dumped down, ignorant, scared and vulnerable, and kept as many as possible into a time loop of a low frequency prison. They intend to continue to trap as many humans as possible before their opportunity closes during the transition.
The human event that will bring us to higher states of consciousness is coming. Many of us incarnate here at this time to experience this big change and now we are starting to remember our true identity.

We have been living in a training simulator that never slept; it had all his drones and minions, which were politicians, oligarchs, actors, banking system, medical system, and religious system which was their stronghold. Death and slavery transcended generations like a virus.

The continuing belief in government, authority and Religion is the most basic setup of the evil controlled Matrix.

Remember, all these organizations like the Federal Reserve, UN, NATO, Red Cross, Trilateral commission, WHO, etc., are just outer circles of this off-world controlled Matrix.
Basically, we have been controlled for our entire existence by a secret global government. Humanity has been under this siege for many thousand years and the losses were heavy, catastrophic, because humans were being assimilated.

When we investigate the endless conspiracies currently being revealed, the deeper we go the closer we get to the core of betrayal of humanity.

These minions have had technology that we are not aware of.

Google, Google homepage, YouTube, Facebook, Yahoo, Bing, these tech companies are being manipulated in ways we cannot counteract and they are using methods that do not leave a paper trail for to be traced, that way we cannot find out how they are manipulating people's opinions.

Dr. Robert Epstein, PhD has done research on these companies, especially Google's ability to control public policy, swing elections and brainwash our children.

The methods Google uses are ephemeral and leave no trail behind, making it very difficult to track and prove if they're using humans as pawns, manipulating us via ways that we can't counteract.

The first step to breaking free from Google's dictatorship is recognizing that the manipulation is occurring; the next step involves consciously opting out of it, as much as possible by protecting your privacy online.

Humanity has been deprived of knowledge through frequency control by the true architects of the old Matrix.
We thought we were alone, but we were never alone, they could read our mind, they were always there, strategically placed always.

Many awakened humans can see currently how the truth is revealed more and is being disclosed at an accelerated rate.
It is now time for change; even though the hardest time is closer to the finale, we must endure. We have thousands of years of real history to recover. Now we need to heal.

"Every record has been destroyed or falsified, every book rewritten, every picture has been repainted, every statue and street building has been renamed, and every date has been altered. And the process is continuing day by day and minute by minute. History has stopped. Nothing exists except an endless present in which the "Party is always right."

~~George Orwell, 1984~~

Last year 2021, we were given a new matrix. The old one was deleted by the benevolent Galactic which has helped and continues to clean out Mother Earth.

But the Evil ones have reinstated the 3D matrix in the beginning of June 2022. Even though we are getting new energy into the new Matrix, now we must start to be aware and manifest positive thoughts and energy. Because there is still too much negativity and evil on Earth, this evil is being generated daily and it is keeping us trapped in parts of the old 3D matrix.

What we are manifesting now is critical for our future and the future of Mother Earth. We are all participants with our intentions and we must anchor our good intentions so that actually we will be manifesting positive thoughts. We must finally free the divine Feminine Energy. Deeper the energy goes, the deeper the sacrifice. We must let go of evil and the forces of dominion and control.

—◁·⫷◈⫸·▷—

January 19, 2023 Teri Wade, www.disclosurenews.it/the-new-matrix-teri-wide/

The Matrix is a grid system that provides the Light required to maintain a physical representation of consciousness through the Human Body.

The Matrix on Earth was hijacked by those who have dominated us for millenia through frequency technology and changes to our body through DNA manipulation.

This new matrix that we are entering in is allowing the human body to ascend physically to a higher dimensional reality.

Light workers and Galactic intervention are bringing in this new grid while the old one slowly dissolves.

Humans that are on the Ascension path have been tuning into this new Matrix. Basically, this new Matrix will have better firewalls so it's less likely to get hacked.

⊐∙⟨⊹⟩⟨⊹⟩∙⊏

Here below is some hidden History that is starting to reappear.
It was written By Lev on disclosurenews.it

In 109,806 BC and 11.008 BC there were catastrophic consequences on Earth.
Two of the tree moons rotating around the Earth collapsed on our planet. This impact created changes in rotation and axis angle of the Earth orbit around the Sun. The poles shifted and started the beginning of the ice age and cosmic night.
(Ps: I learned about the poles shifting from Father Q. Walsh also in 1980s)

This all made the situation of our planet very threatening.

Like other planets in the Universe our Earth is a crystal, its skeleton has the form of an octahedron. It is a crystal that is trying to grow. After the destruction that happened to our planet there were tremendous space wars and the black Archons took over the entire planet.

The Matrix of our 3D Earth logos became asymmetrical resulting in a non-fractal octahedron. So the crystalline of Earth stopped growing with symmetry after the catastrophe of many millennia ago. This led to the weakening of the light vibration coming from the Galactic Ray that brings LIGHT energy to Mother Earth.

This diminution of inflow of LIGHT Creation resulted in the worsening of the climatic social into a downward spiraling of human evolution, life quality and human genetics.

3D Earth was slowly dying.
To stop the Earth's destruction and disintegration, higher Light Hierarchy and Galactic Coalition of many Logos and cosmic races, urgently intervened, they came to Earth and provided mankind with practical help. They developed a plan to save the planet and began to be activated.

144,000 groups of pyramids were built in certain parts of the Planet, the major ones were built in locations of vortices and edges.

The task of the pyramids is to energetically support, reanimate the core of the Earth form (which is a crystal), by restoring and infusing it with Cosmic Life Energy. This brings supports and strengthens the planet structure, stopping the critical decay.

200 pyramids in China in the Canary Islands are pre-diluvial.
20.000 pyramids have been built in India, some in the last 10 years.
In Russia many pyramids are still being built in the present day. At the pyramid in Bosnia, discovered only a few years ago, a leaf was found stuck in the stone layers of the pyramid and the carbon 14 date resulted in being 38.000 years old.
They also build hundreds of cities in the form of stars, fortresses, bastions, forts and walls.

Pyramids in Egypt

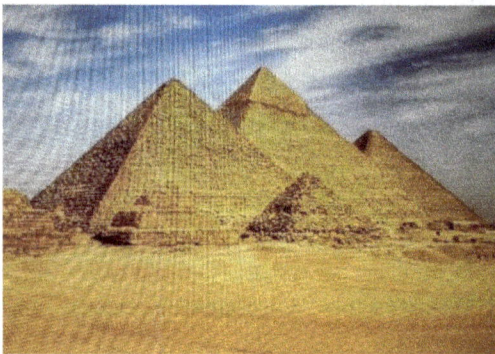

Pyramids in South America

Pyramids in Russia

HIDDEN SECRETS OF ANTARCTICA! By kayleigh McEnany

"They are hiding a lot about Antarctica! It's not what we're being told!"
Mysterious satellite images of 3 pyramids appeared. In July 2023 was found a group of pyramids uncovered from the melting ice.
One pyramid has a square base and resembles the pyramid of Gisa. The square base is 2 km on each side. The Egyptian pyramids appear as a tiny monument in contrast to the ones in Antarctica.
The base of these pyramids are 10 times larger than the Egyptian one which has a base of 230 meters.

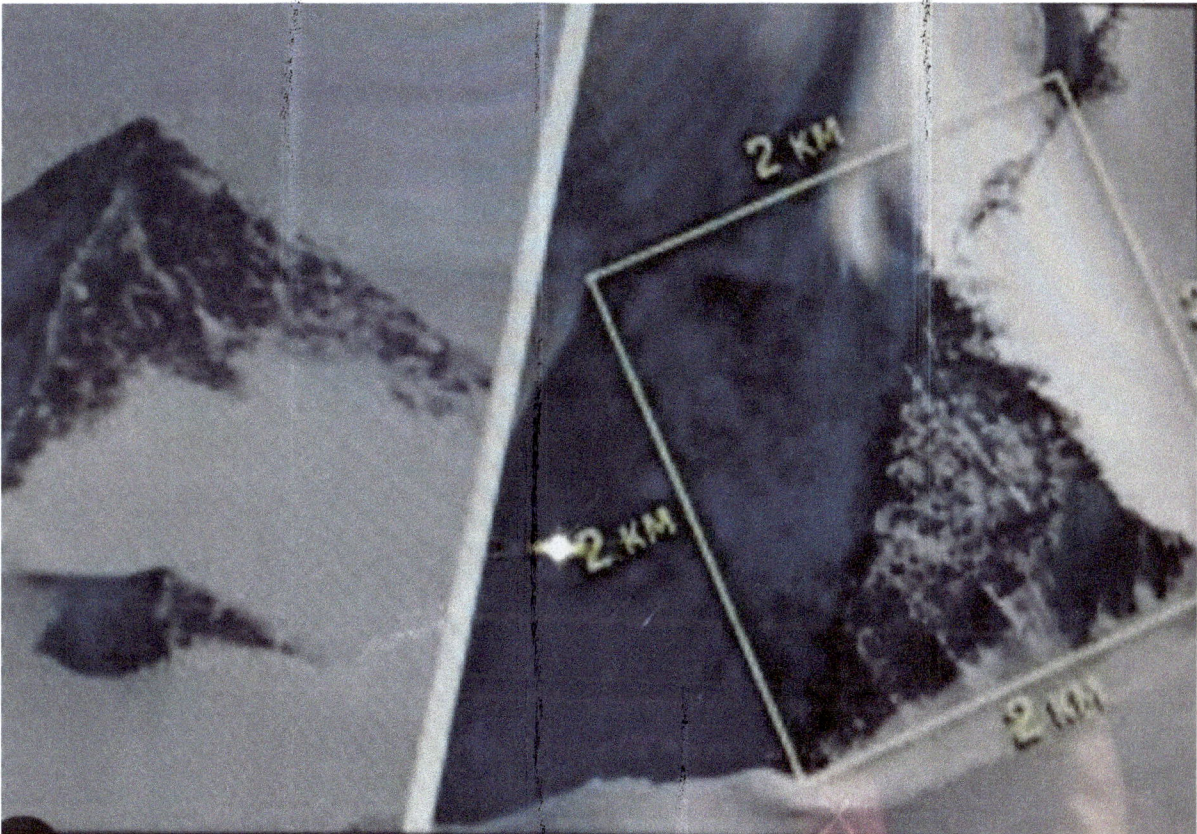

This is a massive structure covered in ice, and is potentially 4000 thousand feet high.
Go on Google Earth and see for yourselves.
//bestnewshere.com/braking-kaylegh-mercenary-drops-bombs-theyre-hiding-a-lot-about-antarctica-not-what-were-told/

Here are some of the star forts and cities

Fort McHenry, Baltimore, Maryland, US

Palmanova, Italy

A world network of "stellar constructions" were built to stop the critical decay. Genuine "stars" objects are projections of specific space Logos and are energy donors.

Bourtange, Netherlands

Almeida, Portugal

According to Lev:

Different configurations of the "stars" correspond to different planetary Logos. All authentic "stars" are located strictly at the points of the Earth crystal grid's nodes.

As a result of the decay, the Earth body slowed down and almost stopped. The nodal points of the crystal grid are similar to the acupuncture points of the human body.

In the center of the star structures, about the size of a square meter, the energy field is completely different then the outside of it. By clairvoyance, one can see and feel an energy vortex, a channel, like a tornado, directly from top to bottom, from the outside to inner Earth.

Different configurations of "Stars" (4-beam, 8-beam, 12-beam) correspond to different types of planetary Logos.

Pyramids' system and stellar network were activated at the same time- about 10,500 BC. That is, before the deep cosmic Night and full capture of the Earth by the Black Archons.

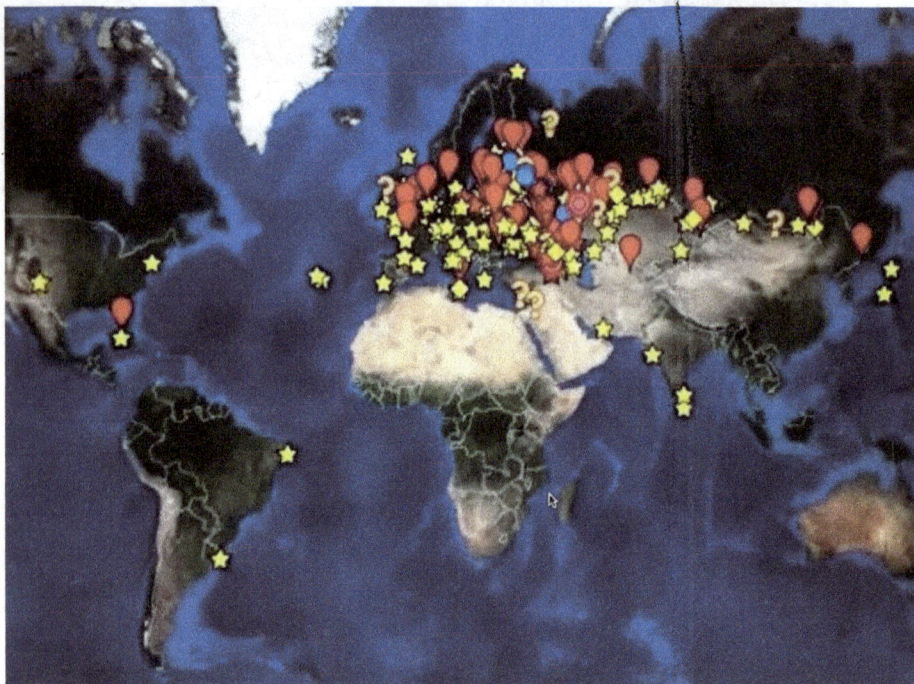

Stellar Network Map

For more details go to: https://www.disclosurenews.it/operation-stellar-network-part-1-lev/
.

Originally we were created intelligent, strong, loving, perfect beings and free with the right to choose.

For many millennia of the past, our planet was ruled during Kali Yuga, by alien dark civilizations.

During the Galactic winter, also called the era of darkness, there was a worsening of the climatic, social conditions on the planet and created a downward spiral of human evolution and quality of life. The evil activities of the rulers were aimed to decompose people and harvest low-frequency crops. There was a loss of spiritual values, love for others and a general economic and spiritual degradation reigning on Earth.

We are at the beginning of the age of Aquarius. The quantum waves of CREATOR LIGHT are being showered on us now, and are bringing SOURCE energy of Love, Purity and Truth.

With our awakening we begin to dispel darkness, and the power of our Planet is passed to the Light Hierarchs.

There is going to be no more a materialistic system, but a spiritual system.
Many Higher Light Hierarchs, Ascended Masters, Curators, Galactic Committee and friendly space species, are working with the Light warriors and Lightworkers here on Earth and are assisting us with Love and Joy to awaken us.

They are connecting with us spiritually and telepathically by downloading into our consciousness and subconscious, good information and knowledge and supporting us with protection and security.

Now the rulers can no longer keep people in obedience like they used to do.

Below I am summarizing a part of an article from Lev:

The function of any Matrix, as well as any field of reality, is made to exchange energy information. We are giving and receiving energies. These exchanges are

1) Equally balanced.
2) Receives more than giving.
3) Gives more than receiving.

The era of Kali Yuga and Cosmic Night was very much energy deficient. The algorithm was written in the 3D Matrix by the evil ones.

If that wasn't enough, the energy was also siphoned by the parasitic-Vampiric system through the portals opened by the Black and Gray Archons and their evil cosmic friends.

Recently the 5D Matrix was created from scratch on the 3D Earth. It was tested and had been firmly installed, now is working, but is not yet in full force.

This new matrix is gradually increasing its potential because it has been careful not to harm any human being with too fast an increase of 5D energy waves coming from the Central Sun of the Galaxy.

The old 3D matrix, which is under the clutches of a huge Hierarchy of the Dark Forces, continues to accumulate and radiate big amounts of negativity of the people and by the war in Syria, Yemen, Ethiopia, Guinea, Mali, Burkina Faso, Ukraine and other countries. These wars have shot off important nodal codes of the crystal lattice of Earth. This caused the reduction of vibrations and overall potential of the general energy of Mother Earth.

In the subtle plane there is a big fight about the two matrix structures. The 3D one continues to work, even though not at maximum capacity as it was before.

Currently, energy and 3D Matrix software are being removed from the Earth's field and new ones are being uploaded in the 5D Matrix.

A big number of the crystal grid's cells of the Earth are not working because they are located under cities, buildings, factories, landfills of any type of waste, an incredible amount of toxic burial grounds and some others are also under nuclear reactors.

All of this is further diminishing the energy of the 5D waves that should be going to the core of Mother Earth in order for the planet to receive 100% of the energy that is needed to the Crystal lattice.

The Co-creators and Higher Light Hierarchs are using bio-cells of the Light warriors. Lightworkers, Starseeds and others, these are the ones who were able to transform their subtle bodies into radiant and multidimensional beings by intensive inner work.

They are the ones that today are receiving large amounts of 5D Light and redirect it to the Earth's core. These wonderful heroes are doing the work in place of those missing parts of dislodged elements of the crystal lattice that doesn't work.

Many ordinary humans are also participating by decision of their Higher Self even if they don't understand or know about it. Portions of new Light frequencies pass through them in waves, for several hours a day.

The main gateway in a human body are the Heart Center, The Crown Chakra, and the Sushumna, which is an energy channel that runs along the center of the spinal column passing through the seven Chakras.

This way around the clock absorption of 5D energies by the Earth's core is achieved. It is very hard work but very necessary, to keep our living planet going.

Our Earth is alive and it is in the middle between youth and adulthood. It is a crystal that has started to grow again. Actually it has accelerated the growth of its physical size. With the increase of the strong vibrations, Earth is changing and we are changing with it.

How fast we are going to transmute depends on the course of WWIII. This war is not only happening on Earth but it is also going on in our local universe. According to Lev the participants are the same as they were during the last war 50,000 years ago in Atlantis.

On June 15, 2022, the evil Dark ones attacked the planetary abodes of our Stellar Ancestors. And with the help of black magic ops attempts were made to kill leaders of states and destroy Logos and egregores of those countries. They also attacked Light warriors which are hunted and tracked by their luminosity.

With the combined efforts of the Light Forces, the assault was defeated, even though it caused a lot of losses. The present threat is over, but the situation is tense and it could escalate. It is important to be alert and ready for more surprises.

Let's stay awake as we traverse the valley of the shadow and death, on the other side of destruction is Creation, which awaits the deeper illumination of higher consciousness that we have worked so hard to achieve. May the spirit of Christ walk with you, in Purity, Kindness, Discipline, Patience, Diligence, and Humility!

The event is our return to consciousness.

I found this writing here below many years ago and I wrote it down, I don't remember where I got it from. I repeat it often to make myself strong and defeat fear and evil.
If many people repeat this, it will weaken the evil ones even more.

1) We do not consent to the porn industry that has been the breeding ground for sexual perversion. They have no power over me and all of us.

2) We revoke, remove and dissolve all banking family trusts that use our name as numbers for energy harvesting.

3) We are eliminating the consent to etheric grid broadcasting over public airways, television, internet, radio and other means or methods to propagate propaganda for mass consumption.

4) We are revoking all contracts with the sacred geometry satellite broadcasting system that service the few that want to dominate and control humanity.

5) We revoke all spiritual contracts secretly interwoven into each and every law, past, present and future.

6) We revoke and do not consent to the use of sacred geometry Cities, infrastructures, linguistics or artwork for the purpose of domination and control.

7) We do not consent to the rule of privilege few.

8) We do not consent to timeline wars.

9) We do not consent to mass propaganda.

10) We do not consent to war based industrial complexes.

11) We do not consent to our life force being used for any purpose that doesn't heal Earth Mother and promote unity Consciousness among all sentient beings.

November 20, 2023,   Written Report from Benjamin Fulford's website

The geopolitical situation is in the middle of a collapse of the USSR-type situation. This will lead to many countries disappearing from the map in their current form. Countries likely to disappear include Israel, the United States, Ukraine and Poland. This comes as Rockefeller stooge president Joe Biden is set to sign surrender documents in San Francisco this week to the greater planetary liberation alliance, Western White hat and Asian secret society sources say.

What is happening is that a centuries-old Satanic plan to use three world wars to turn the planet into a giant slave plantation has ended in failure. The Khazarian Mafia hoped to use this plan to rule the world from Israel and greater Khazaria (Ukraine Kazakhstan etc.). Instead, as we shall see below, the KM have been decisively defeated in the Ukraine and will soon be in Israel.

"Ukraine is an essential piece that we cannot afford to lose on the geopolitical chessboard…our current path…means that our global order is dead on its feet," admits Nathaniel Rothschild who has taken over the KM now that much of the Octagon group has been neutralized.

The defeat of this plan means many borders may return to a situation similar to what existed before the KM engineered World War I, the sources say.

In other words, Germany will return to something of its First Reich borders, the Austro-Hungarian Empire may re-emerge as a republic and the Turkish Ottoman Empire may once again turn Judea (misleadingly called Palestine or Israel) into a protectorate. Needless to say the rights and autonomy of the Poles, Jews and Judeans ("Palestinians"), etc. would be protected under such a scenario.

However, the changes may be more far-reaching than that because the satanic UNITED STATES OF AMERICA CORPORATION is set to be dismantled as a result of the victory of the American people in their second revolution. The Republic of the United States of North America is likely to emerge as a replacement for the corporation when all the dust settles. This may all sound far-fetched but facts in the real world make this the most likely scenario.

Let us start with the situation surrounding IS (ISIS the moon goddess) RA (the Egyptian sun god) EL (the creator).

The Jews know their own government attacked them to justify war in the Gaza Strip. A leaked video from the Israeli Air Force shows that it was the Israeli Apache helicopters that bombed its citizens during the "Delirium" festival on October 7, not Hamas!
Also, now even Jewish holocaust survivors are getting arrested for antisemitism [anti-satanism].

No wonder Haaretz, a major mainstream Israeli Newspaper said:
Prime Minister Benjamin Netanyahu, who is meant to be leading the country, is a haunted politician facing the end of his career, with the present troubles compounding

the serious criminal entanglement into which he maneuvered himself with his own
hands. Netanyahu does not enjoy the public's confidence, and most of his efforts are
invested in his personal survival.

Last week leaders of all the Muslim countries gathered in response to the Satanic massacre of Judeans in Gaza ordered by the Satanist Netanyahu.
At the meeting of the Organization of Islamic Cooperation (IOC) in Saudi Arabia last weekend a consensus was reached to end Israeli mischief once and for all. The IOC countries agreed they would not be fooled by the KM into starting World War III. Instead, they will follow the playbook of the neighbors of the original Khazaria.

The IOC will issue an ultimatum to Israel and its' KM overlords to stop their criminal and antisocial behavior around the world. They will threaten to attack Israel with an army of over 5 million that outnumbers them by 10 to 1 unless they arrest the war criminals in their government and subject themselves to Turkish guardianship.

This is why Tass reports Turkish President Recep Tayyip Erdogan is planning a "global initiative" to resolve the Gaza crisis.

https://tass.com/world/1703703
Also, look at how Rockefeller slave Anthony Blinken gets the cold shoulder in Turkey as he tries to forestall this move. The current US regime has no credibility in the region. Furthermore, any nuclear blackmail attempt by the Israelis will be countered by Pakistan's nuclear arsenal. The Chinese and Russians also support such a plan.

No matter what though, this is not going to turn into the KM's long-planned Gog (the G7) versus Magog (The SCO) all-out nuclear war scenario to kill 90% of humanity. The US, Chinese and Russian militaries will not go along. The US military is no longer under the control of the KM, Pentagon sources explain. Even if compromised leaders try to order such a scenario, 72% of Americans say they will not support their armed forces in the event of a major war. The military rank and file is with the American people and not the

KM.

https://www.newsweek.com/american-military-recruitment-problems-public-apathy-1842449

Take a look at how CNN tries to suppress real US military views by cutting off this soldier. In any case, we are getting reports the US mainland itself is now under attack, meaning no troops would be available to fight in the Middle East. The most dramatic evidence of this is coming from Los Angeles. "Every night around 2 AM US military troops arrive by truck along the main boulevards of Long Beach California. The troops are going underground and loud explosions that are not earthquakes are being heard. When I asked a soldier at the cafeteria of the port facility who they were fighting he said they were fighting.

CHAPTER 11

—◁◦⬦◦▷—

THE PRECESSION OF THE EQUINOX OUR ENTIRE DNA RECONNECTED

In 2012 our Solar system ended a cycle called the Precession of the Equinox.

This brought us to a new beginning cycle that started a new and completely different reality on planet Earth.

Our Solar system is a part of a much larger system called the Pleiades.
The Pleiades are a group of 7 stars systems also called the Seven Sisters, which are rotating around their bigger and brighter star Alcyone. Our Sun follows the seven stars through the cosmos.

Our Earth completes a rotation around the Sun in 365 days. Our Solar system makes a revolution around the star Alcyone in 25,630 years, which is called the Precession of the Equinox. Alcyone makes a revolution around the Great Central Sun of our Galaxy in 225 millions years.

We have witnessed the end of the age of Pisces, which was about control and oppression, low frequency reality, and darkness. We are now in the beginning of a new age, the Age of Aquarius, which will bring us to the Golden Age.

We are now showered by an ocean of intense Light with high frequency waves that are coming from the central Sun of the Galaxy. These WAVES of LIGHT reach the star Alcyone that eventually sends it to our Sun which is sending to our Earth "X-Class" Flares through Coronal Mass Ejections of extremely profound intensity. Scientific research has discovered that it now takes approximately eight minutes for this LIGHT to arrive on our planet.

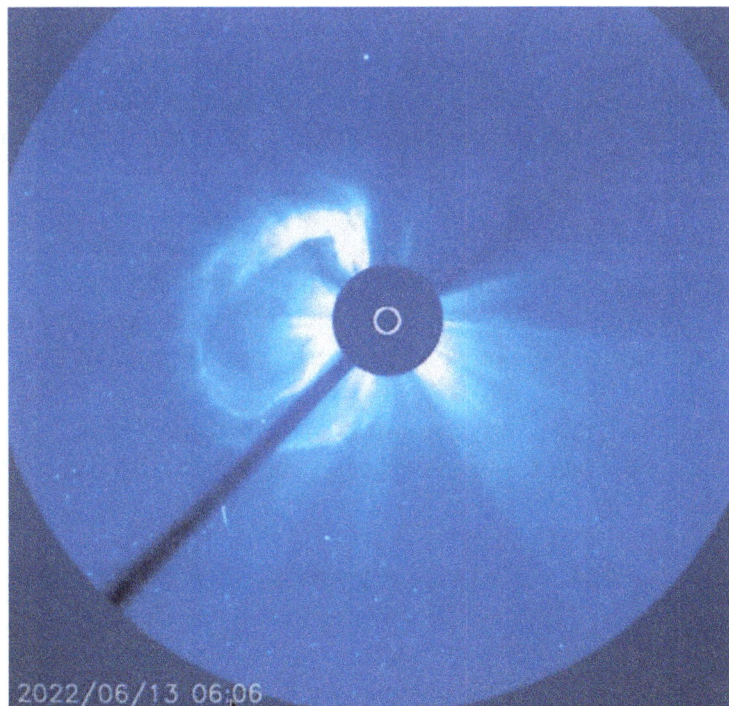

Coronal Mass Ejection 13 June 2022

All of this cosmic transition activity causes our planet to vibrate faster on its axis as it receives brighter photons of LIGHT.

In the last three months of 2020, the Karma Lords had cleared all the bad karma from our bodies and from the Earth to get us ready to receive a much stronger LIGHT from Creation. This cleansing of Karma was needed or we would have perished when the new stronger Waves of LIGHT started coming in January 2021. The cosmic night was over!

As a consequence everything on Earth including our bodies are now vibrating faster, even our heart rhythms are quicker, and all these heat surges are changing us, while we get accustomed to assimilate SOURCE LIGHT FREQUENCY.

Even though the corrupted mainstream media doesn't tell us, and NASA is keeping it a secret, I have known that since January 1, 2021 we have been showered with very strong LIGHT waves coming from the central Sun of our Galaxy. They are much stronger than we ever had before.

That LIGHT has removed the veil that prevented us from being connected with our GOD SOURCE. This incredible LIGHT is pouring on us at an ever faster and stronger pace than ever before, and it is also changing our biology. Everyone is affected by this cosmic current of energy which is allowing us to have a higher connection with our Higher Self.

Our bodies are transformed from carbon "based" beings into crystalline "based," and this new LIGHT is healing and upgrading us. We must surrender to this process. Do not resist it.

Everyone is experiencing this process at a different pace depending on how it was planned before the soul incarnated into a body. And also for how long we have been detoxing our body. Because of the artificial programming that we have been subjected to. The Deep State has been using chemicals in our food, GMO, electromagnetic frequencies, political and religious manipulation of our mind and body.

We don't live any more in a 3rd dimensional space. Mother Earth is vibrating at higher frequencies of 4th and 5th dimensions. All of us are being slowly acquainted with this new vibration including the animals and everything else that lives on the planet even the plants. Some types of plants are disappearing while other plants are blooming profusely.

For 20 years I had a Lily plant that always bloomed in the month of May, but during the last 3 years it started to bloom in the months of January and February and continues to bloom all the way through the month of May.

This morning December 5--2023, while I was watering all my plants I saw one of the Lily plants had the first white flower buds. From last year to this year it anticipated an entire month. This shows me how stronger the quantum waves of Creator LIGHT are coming to us now, it is showering us with Light and it is bringing healing to our body and higher consciousness to our heart and mind.

Photo of my Lily in March 2021

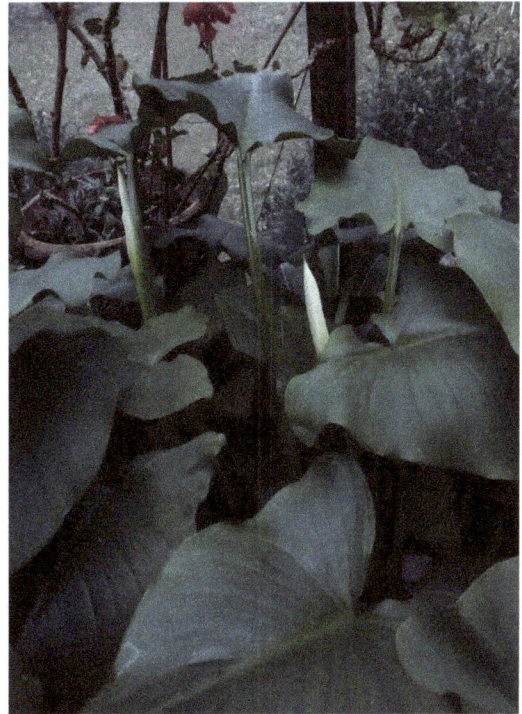

First buds appearing in February 2022

First buds appearing during the first week of December 2023

There is even more LIGHT coming in and my Lily plants are blooming even earlier than ever. From March 2022 there was a crescendo of LIGHT, to February 2023 and now December 2023 it is a whole month earlier than last year.

All Creation is going through an incredible transformation.
As we acclimate our bodies to this new LIGHT and vibrational frequency, every one of our body cells is learning to vibrate at a higher rate. As the cosmic forces are increasing we are advancing in stages of transmutation. There will be a transformation of our spiritual and energetic structure. At the first stage, purification of the body is needed, including cleansing of our internal organs.

To resist the change and cling to a reality that has controlled the masses for far too long is no longer possible. Our Spiritual evolution is unstoppable and the system that has controlled us and the Earth for thousands of years is breaking down and generating chaos.

The end of the cycle of chaos will bring only peace for humanity.

We are clearing out, dissipating all the dense 3D energies from our body and cellular memories. Change is constant!

SOURCE LIGHT is making us morph into crystalline beings. The 10 strands of junk DNA, so called by the scientists (because they said, it is dormant), in reality it was cut off, now it is being reconnected, calibrated and will return to the way it used to be when we were perfectly created beings. The pulsar codes are activated by the incoming LIGHT and are changing the dormant DNA.

The Evil ones who had taken over the Earth and subjugated humanity many thousands of years ago, cut these 10 DNA strands because they were interdimensional strands that used to interact with the other two biological strands of DNA that we still have functioning. Without all the 12 DNA strands working together we are unable to connect to other dimensions and rise to higher densities.
Our evolution would not progress if the 10 strands of DNA continued to stay severed.

Over the years and millenia, humanity has been living with 10 DNA strands being cut so we were enabled to communicate with the other dimensions. It kept us in a very low vibration and isolated from God Source. Over the years things got worse, toxins in food, in water, air mixed with poison, vaccines, electromagnetic fields and the constant pushing exposure to dangerous chemicals into humanity's lives and bodies, are all things that lower our energy and strength. This is how the evil ones keep us from evolving. But now all of our DNA and cellular records are being cleared, purified and enlightened with the Powerful LIGHT from the GREAT CENTRAL SUN.

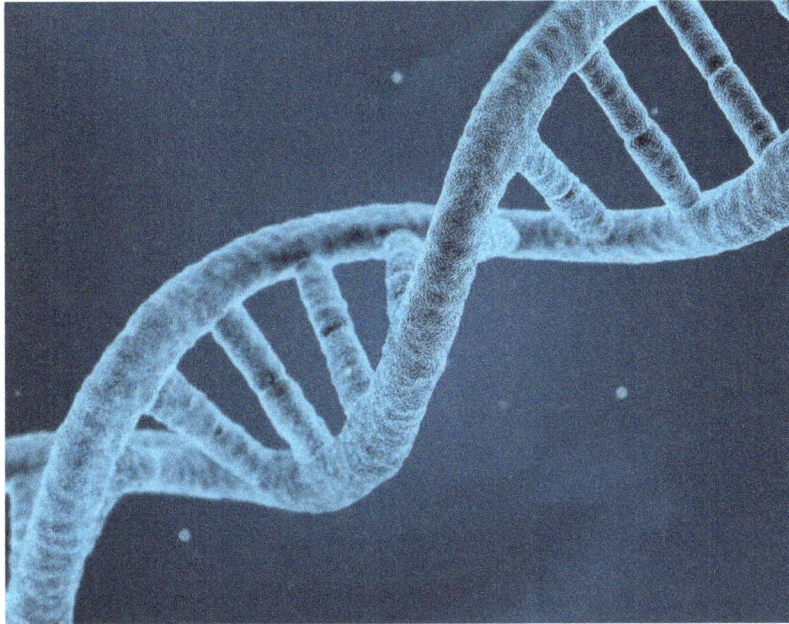

Our biological DNA

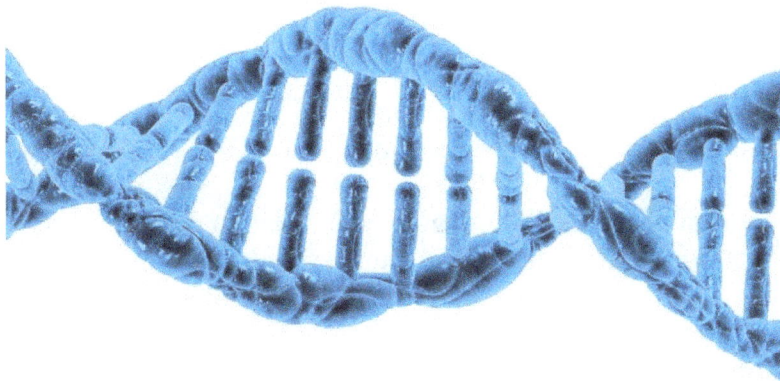

10 interdimensional DNA strands that were cut

This DIVINE LIGHT from SOURCE flooding into us more and more each day is rewiring our DNA so that all 12 primary strands of DNA function as nano-sized strands of optical fibers. They will eventually re-link and function together with the 12 petal multidimensional chakra portals, allowing the Akashic light to flow in and out of the human body.

In these days of accelerated waves of Quantum Light energy, many of us have been going through a profound transformation of our physical body. This is a process that is unfolding in all those humans who did conscious inner work, perfected themselves and their soul, and learned to innerstand the universal laws and live in accordance with those laws.

Those people who voluntarily and consciously cleansed their body of limiting beliefs, and negative toxic emotions, through their actions, and learned to love and accept people and the world the way they are, eventually will access consciousness and they will feel as though they finally came home from a long journey.

CREATION LIGHT WAVES are now coming to us in a higher and faster way. Our transformation continues on a more accelerated massive scale, depending on how much purity a person has achieved so far, it will make them ready for the next stage to higher consciousness and pass it easily or not.

Beyond the chaos and darkness we will see all with the LIGHT that we shine!

Spiritual practices both ancient and modern will accelerate the process.
The first stage is the transformation of the physical body. Then the body must go under 3 more phases of transformation before the new wave DNA of multidimensional activation begins.

According to Lev these are the three Phases:
Phase 1 - Restorative (purification-resurrection)
Phase 2 - Chemical transformation (transmutation)
Phase 3 - Code replacement (resurrection- ascension)

As we go through the first phase we go through periodic pain in the endocrine glands and chakras. This will reinforce our circulatory and lymphatic system, digestive and excretory system, respiratory system, the heart, and our bones. There will also be changes in our electrical system etc.
Our bodies will begin to rejuvenate.

Here is the link where you can read Lev's extensive article about all of it.
https://www.disclosurenews.it/our-wyrd-from-3d-to-4d-and-5d-lev/

Our body is being switched from material to physical LIGHT BODIES.

A new evolutionary corridor is being opened in front of us. The frequency of our bodies is going up, higher and higher, as we are going to step into the 4th and 5th dimension.
In these new dimensions different laws apply.

The spiritual transformation in a physical form has never happened before in our Universe. For this reason all the Galactic in the universe are now observing, monitoring and helping us through this process, with excitement and anticipation.

We must allow and surrender to this change. We must make certain that we are engaging transformational energy and not destructive energies.

Each person goes through this ascension process in its own individual way, depending on the preparation work that was done throughout many past lifetimes, good, bad and ugly.

Duality will vanish.
The unseen will be seen!
The unknown will be known!

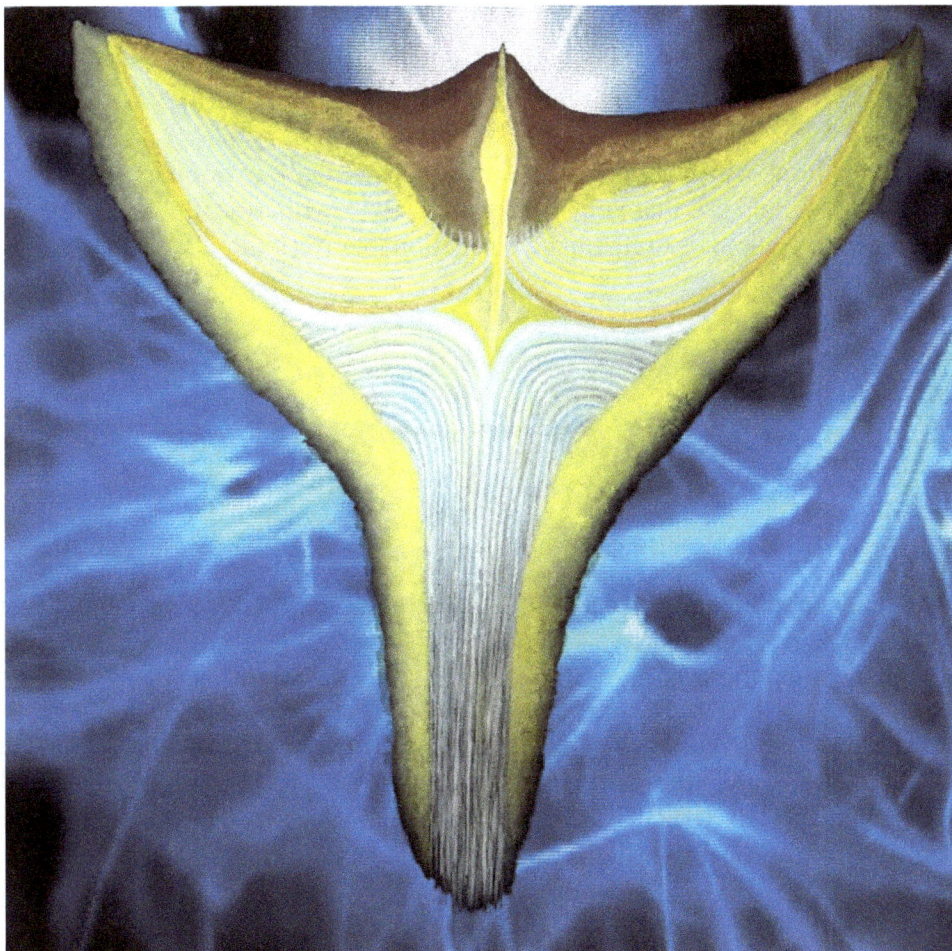

We are like the Phoenix which rises out of the ashes of the blaze.

This is a writing from the Pleiadian collective!

Each moment that you breathe is a gift!
Every day when you awake from slumber in your bed, find the love and gratitude that you want to express!
Before you enter your world of form, be still and quiet and know you are Spirit!
You are the essence of God that has infinitely existed as all life!

Through Galaxy and dreams you were assisted to arrive in this dream with others!
With the Galactic forces, Angelic Realms, Fairies and Guides, you are not alone!

You will be free of darkness and find comfort within, until the moment when LIGHT covers the Earth!
To all, you are so Loved and so worthy to be here!
Be a gift to yourself on this day!

In every moment know that a path is already open for you!
To those that feel bound, turn to the LIGHT of GOD and know that everything is working out for you!
Set your intention to be LOVE as you walk with Souls in disguise of humans!
Know that beyond the Veil that will soon fall, they are your family and friends!

Choose thoughts that bring a higher perspective and choose LOVE over Fear!
I cover you with LIGHT and ask that you share your LIGHT with others!
This Season, This moment knows that you are perfection manifested!
Remember who you are!

In love I remain with you!
EN EEKE MAI EA!

———◦∙❮❙◦❙❯∙◦———

Dr Schavi M.Ali

If more people understood the importance of tuning into SOURCE for guidance, our world would quickly become a better place.
If more people understood the messages of the cosmos, this too would be ideal.

If more people knew the truths of history rather than being locked into indoctrinations, the "Peace Pipe" could be shared in the "Circle of Humanity". We would be more peaceful, healthy, abundant, and elevated to HIGHER KNOWLEDGE as the "norm" rather than having to continually solve serious problems which seem to escalate daily.

———◦∙❮❙◦❙❯∙◦———

One of the last things that I remember is Father Walsh saying that he believed that the world would be saved by the Women. And in these dark difficult days I see and read more and more of women researching for the truth, writing books, blogs and digging into the government's archives exposing all the lies and evil that was accumulated over the past years and millennia.

The women are presenting videos with others like this ones:

Host: Stephanie Loricchio- Guests: Robyn O'Brien, Jim Gale, Kate Tietje, Ami Bohn.

FOOD SECURITY IS NATIONAL SECURITY
https://live.childrenshealthdefense.org/chd-tv/shows/good-morning-chd/food-security-is-national
-security/?utm_source=email&utm_medium=blackbaud&utm_campaign=chdtv&eType=EmailB
lastContent&eId=d156ac9b-d48f-4a75-adb2-bf3f761cd2fb

LESS PHARMA, MORE REGENERATIVE FARMS
https://live.childrenshealthdefense.org/chd-tv/shows/good-morning-chd/less-pharma-more-regen
erative-farms/

Corey Lynn September 29-2022 COREY'S DIGS

GLOBAL, LAW & ORDER, U.S.
LAUNDERING WITH IMMUNITY: THE CONTROL FRAMEWORK-PART 1
https://www.coreysdigs.com/u-s/laundering-with-immunity-the-control-framework-part-1/

Funding the Control Grid Part 2: The Psychological Framework
January 27-2023
https://www.coreysdigs.com/health-science/funding-the-control-grid-part-2-the-psychological-fr
amework/

HEALTH & SCIENCE, U.S.

The Global Landscape on Vaccine ID Passports Part 4: BLOCKCHAINED
August 26-2021 What the world economic Forum predicts for the world in 2030:
Normalization of QR Codes To Access Your Data, Your DNA, and Your BODY
https://www.coreysdigs.com/technology/the-global-landscape-on-vaccine-id-passports-part-4-bl
ckchained/

REALLYGRACEFUL, BOOK "THE DEEP STATE ENCYCLOPEDIA"
Exposing cabal's playbook.
March 2023

ISRAEL: LEADING THE WAY TO TRANSHUMANISM

**Ilana Rachel Daniel reports on transhumanism and the continued push for centralizing
control on "Good Morning CHD". She sits down with Catherine Austin Fitts to dive deeper
into these topics and their implications on our lives. Watch the episode on ChD.TV!
April 19, 2023**

https://live.childrenshealthdefense.org/chd-tv/shows/good-morning-chd/israel-leading-the-way-to-transhumanism/

Michelle Stiles reviews some of the basic tools of propaganda, in her book:
"One Idea To Rule Them All: Reverse Engineering American Propaganda
on Amazon

"In the age of universal deceit, telling the truth is a revolutionary act."

THE WORLD IS WAKING UP! THE TYRANNY WILL END!
They are few, we are many!

CHAPTER 12

SOLAR STORMS

A Solar storm is a massive explosion in the Sun's atmosphere. These activities of the Sun are also called Solar flares, Solar winds, Coronal mass ejections. Throughout the month of June 2022 the activity has been at the highest levels of intensity and it gets higher every week. SOURCE LIGHT is pouring down on Earth. On June 26, 2022 the cosmic energy was so dynamic coming from SOURCE LIGHT, that the technological equipment on Earth could not adequately register the true strength of the incoming LIGHT!

Every 25,630 years our Sun and the Great Sun Alcyone align and the connection to SOURCE frequency is far greater. Major changes start to happen on Earth which is now united with chaos and turmoil. SOURCE LIGHT frequency starts to stimulate the clearing of cellular records, cleansing of old programming, which gives enormous, powerful changes to humanity.

These flashes of LIGHT contain protons, electrons, carbon dioxide gasses and immense amounts of plasma and other substances. All these extremely powerful LIGHT waves cleanse, clear, regenerate and revitalize the crystallization of carbon molecules. Studies have found a direct connection between solar storms and our human biology. We are physically, mentally and emotionally altered by the electromagnetic activity of the Sun.

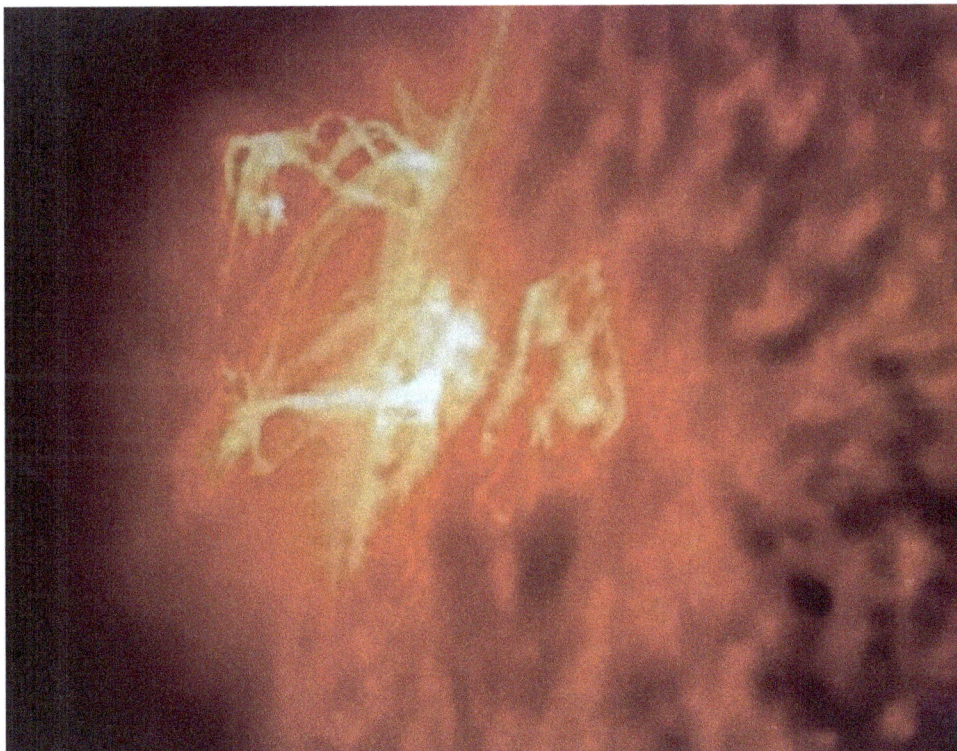

Solar Flare 10 June 2022

These strong LIGHT waves, coming from the Great Central Sun, impact our Sun and rebound on our Earth. The powerful SOURCE LIGHT and intense energy is very beneficial to our cells, organs, bones, and it is healing because it is affecting every part of our physical body, every muscle, gland, organs, nerve, endocrine system and chakras.

165

The Light Activation Symptoms (LAS) also called "Ascension Symptoms" will not spare anyone.

Intense LIGHT is not only making our physical bodies change, but also our emotional and mental symptoms. The upgrades do not come online as soon as the LIGHT enters our physical body, but we begin to feel the symptoms some time later, as the body readjusts to the new LIGHT. Change is painful and we experience stress and pain.

Every time a new wave of strong Source Light comes to Earth, it affects us in different ways: headaches, dizziness, sinus and throat congestion, nausea, diarrhea, ringing of the ear, joint pain, muscle pain, anxiety, palpitations, vivid dreams, depression and lack of normal sleeping patterns.

This is because of a shift of DNA in the cellular atomic and subatomic structure, tiredness, feeling burn-out, extremely exhausted and drained, general discomfort, feeling lethargic etc. Each of us experience different symptoms, depending on different imbalances that need to be cleared out.

All of this LIGHT awakens deep seated and hidden emotions which must be released in order to heal. This is "awakening" dormant records as well as creating new ones, including "downloads" of higher knowledge.

The human beings who elevate in consciousness become a laboratory of "Spiritual Science" and a book of higher knowledge. Some of us are becoming empathetic; others are becoming highly intuitive from being constantly connected to messages from their HIGHER SELF which allows them to be compassionate towards the challenges of others.

Not everyone will experience all of these symptoms, but some of us will experience quite a few of them. I don't run to the doctor because usually the pain goes away on its own. But if the pain is too severe for many days, I get some help.

These waves of energy are showering us with a Golden, Silver Violet flame. This is a powerful Light that makes us grow to higher states of consciousness. The solar storms are now helping humanity to reconnect with SOURCE DIVINE LIGHT. The transformation taking place within us is reflected in the collective, this is by Divine design.

We are the children of the LIGHT!
The best is yet to come!

We arrived here on Earth as a soul, we incarnated in a body, in the end of this experience we will leave as a lighter soul/body.
Nothing matters but SOURCE LIGHT. We are all ONE breathing as many! Every moment that you breathe is a gift!... In a world full of chatter, listen to your own truth within.

Both small and large-scale solar flares; coronal mass ejections with each flare; the growing solar wind currents; the constantly arriving plasma particles of protons, neutrons and electrons. The human physical vessel, emotions and mind are all changed and shifted, and so it is our planet.

Regardless of where people reside, in the northern or southern hemisphere, there are increasing reports of unusual 'triggers/symptoms' coming from healthy people who have not taken the 'vaxx' or 'booster' and age is not a factor. Everyone is "preparing" for the most exciting event humanity has experienced in eons, and it won't just be the arrival of ships from outer galaxies, although that is part of the scenario.

Many ancient books and gurus have been telling us to go within ourselves. There, in the inside of us, in our soul, we will find peace and truth.

Consider this place that you discover within yourself is your home. It is a place where anything you sense you can perceive, it is your true state of being.

When we incarnated on 3D Earth, our soul forgot who it really was.

With each incarnation we have had the impression that we have never existed before and that the death of the body means the end of everything. This resulted in a feeling of complete isolation from our Source of Life. Many of us do not even know of its existence, so strong is their amnesia.

We are immersed in an illusion from which we must awaken.

When we connect with Nature and immerse ourselves in it by observing it closely we feel different, we remember faster and we understand the path of ascent of our soul. Knowledge and wisdom save us in the most difficult conditions of third dimensional frequencies,

We must stop the mind from running wild and in fear. Our inner guidance is our ultimate authority that has only our best interest in mind.

It is now important that we start to think in a new way. Nothing will change without a mindset change.

"If we do what we've always done, we will get what we've always had."

Change is life, there is already a physical process going on in our body at a molecular level. It is important to allow this great transmutation to change us.

"Change is hard at first, messy in the middle and gorgeous at the end"

~~Robin Sharma~~Writer~

LIGHT from both inside and outside our solar system comes to us in various ways of intensity. Are we preparing for a dynamic and dramatic planetary event through hundreds of daily earthquakes?

Today November 30- 2022
There have been 155 medium level "4.0" and "5.0" earthquakes with one strong "5.6" occurring in Iran. The solar wind increased to a whopping "728" km/s and then rapidly decreased to "695" km/s.

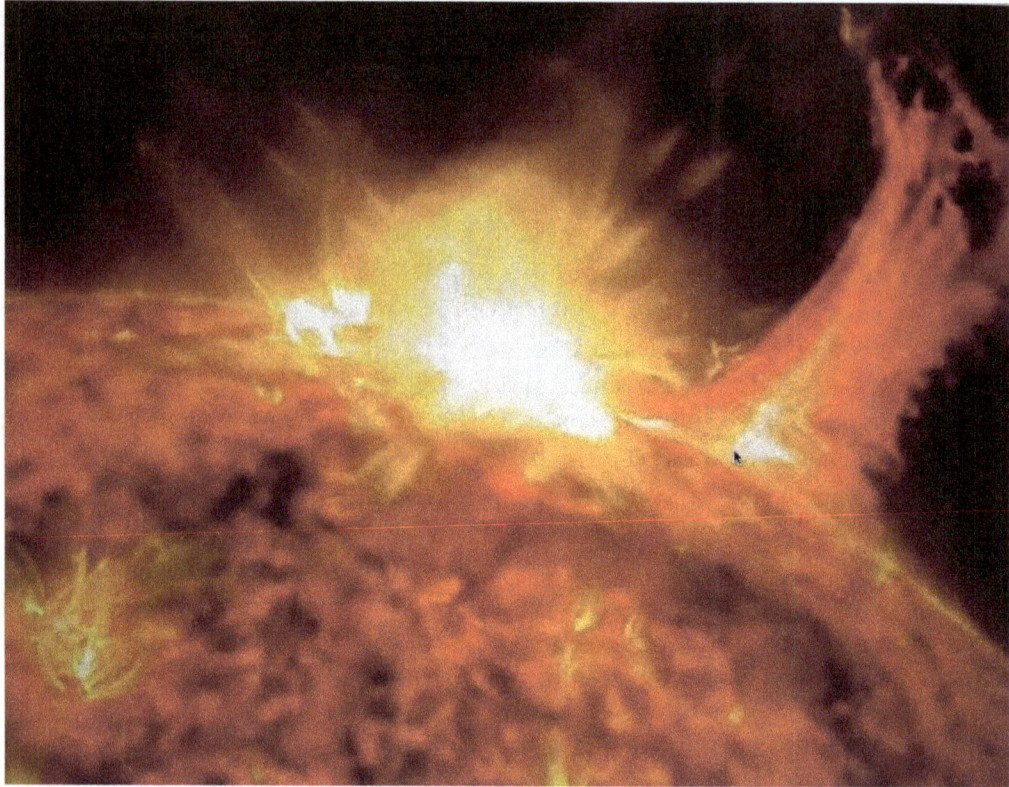

Solar Flare 7 February 2023

Recently I read a new article from Lev title:

HOW WE ARE CHANGING

Some people started to notice a Halo over the head of humans around the world.
During the summer of 2014 a Magnetar appeared in the place of a black hole in the center of the Milky Way.
This Magnetar/star began to glow and pulsate, expelling a substance in form of big and small granules.
At first the astrophysicists heard the sound in a radio wave range, eventually they saw it with the use of instruments.
This silvery halo is caused by the radiation of the Hippocampus, which is the part of our brain that helps the formation of theta rhythms for concentration, and also as to do with the formation of memory and emotions.
It seem that this substance, which they have not given a name yet, brings intelligent plasma, it makes the Blue spot in our brain glow, while is inoculating new programs into us through the subtle plane using the same frequency.
As we go through our lives we do not know that now we use a different multidimensional energy spectrum.
In the old renaissance paintings in the churches and basilica the Saints and highly spiritual people were painted with a Halo above their heads. the Halo didn't appeared over the head of common people. The reason was because in the old days there was a rigid magnetic field around the Earth, and that prevented our body from receiving higher energy information.

Now the frequency of the Earth is completely different, because the Schumann Resonance is much higher. The cells membrane of our body are linked to the frequency of the Earth.

I remember that I found out about the Schumann resonance when I was 17 years old, at the bus stop while waiting to be taken, with other students, to another city where our art school was located. We had conversations about the frequency of the Earth, then the resonance was always at 7.8 Hz. Many years later I learned that this frequency was the low frequency that the Archons were using on Earth to keep us in a low vibration, slavery mode.

Everything has changed now, I remember after 2012 I started to check the frequency more often and realized that it started to fluctuate higher. As time went by, the frequency started to fluctuate even higher, and one day reached 100 Hz.
Then the first time the frequency reached 150 Hz I was very surprised, because I felt different, everyone I talked to told me that felt strange feelings too.

Lev said that the Microbiologists sounded the alarm because according to them our cell membrane could not take a frequency higher than 13 Hz.
It seem though that in 2005 the 13 Hz barrier was crossed successfully by us, and from 2014 on it opened up a corridor of Joy, and conscious creativity, while from 150 Hz and above came the energy of Love and Mercy.

The Magnetar emits powerful codes that make our DNA soft and malleable, this way it is suited for upgrading. It revealed that while before there was a protein under our nails to make it grows, now that protein is found everywhere in our bodies. In the hair, stomach, intestine, in the brain, in the epithelium and the neural network.
It was discovered that it belongs to an ancient code set for regeneration of the human genetic structure. It is like a magic wand that becomes activated and goes to the places where our body needs to be fixed.

Our perception is changing and we are on the road to be the children of the LIGHT reconnecting with our multidimensionality the way we were created originally.
Yes all this changes to transform our body into a crystalline state, at times is painful. The Light Activation Symptoms are making us sometimes feeling depressed, apathetic, feverish, not being able to sleep, having pain in different parts of our body, etc.
One day during the summer of 2022 I felt that my heart had stopped beating, I had non more pulse and I could not breathe.. I thought I was dying. I sat on the floor, however after a minute or little more later everything started working again, but I felt a little dizzy as I got up off the floor after that strange unexplainable thing. It puzzled me, but I went about my day as if nothing had happened, because what I had experienced was inexplicable to me at that time.

I recently discovered while reading Lev's article that this was a brain transition signal, in a short term it reconnects us with multidimensionality. In esotericism it is called the reception of Agni, a Sanskrit word meaning fire.

Here is the link to read the whole article:
https://www.disclosurenews.it/how-we-are-changing-from-3d-to-4d-and-5d-lev/

Be in a position of strength in each moment! Practice and Love are our Great Gifts. Every awakened person is worth hundreds or even thousands of unawakened ones.
When the awakened ones communicate with other like minded beings, they carry a form of public opinion which has a completely different vibration, it is the energy of freedom and truth. They influence the collective consciousness to begin to raise even more positive vibrations, displacing the low vibrations of the 3D world of the Archons.

With each passing day in 2022, the destructive and constructive karmic programs in the mind and heart of all beings are affected and those in 3rd, 4th, and 5th dimension will be eventually completely separated from one another. Those who are still attached to the 3D low frequency, materialistic way of thinking, will remain behind, and the others will move on to the new 4th and 5th dimensional Earth.

The quantum waves are shaking down the Black Archons' power. We are reaching the final act; humanity is finally going to be liberated from all this horrible tyrannical system created by the Black Archons soul eaters.
For millennia evil has corrupted the Earth and humanity. These horrible creatures have changed us by genetically modifying our bodies, and socially engineered our civilization. They keep us slaves and violent in order to disconnect us from the Collective Spirit which unite us to GOD SOURCE.

Humanity is moving to a new level of evolution, even though the Archons are losing this battle now, they are still attempting everything, using all they have in their Dark arsenal, to derail our awakening transition.

We are living in a rare precious moment in the history of our planet, when many natural cycles are converging in a way we have never seen before. This convergence is changing everything.

The old ideas are no longer sustainable. The old things are being dismantled.

So many systems are collapsing: Energy, economy, food, water, defense, communication, and the way we share vital resources.

We are going through a revolution that is so much different than all other revolutions in the past history of humanity.

This breakdown looks like chaos, but this is the way the Universe works. When a system becomes too complex and so difficult to bear, because it is no longer sustainable for the Earth, the humans, the animals and the plants, chaos scrambles everything up with intensity for a while.

To maneuver the pathways through chaos, there must be a shift in awareness. When enough is enough and everything changes, at first we go into denial, then we go through a period of adaptation to the changes, in order to become a better version of ourselves and to arrive at higher states of consciousness. Then everything becomes normal, we become adapted to the new way.

By focusing our attention on the maintenance of our form in a state of purification we will maximize our chances to complete the process of accessing higher states of consciousness so we can send the best vibrations to transcend the limitations of a system, which is completely collapsing.
In a world full of chatter, listen to your own truth within.
The pathway is not outside of us, it is inward for all beings. Let's make our internal life more important than the external life and accomplish the goal to enter higher realms.

This DIVINE LIGHT and our inner work will eventually bring us to higher dimensions.

As we start to awaken to the truth, some of us don't really want to know what happened to ourselves and to the planet in general, because it is too painful to see the truth. Others are too busy making a living and worrying about paying the bills because of the rising higher cost of living, food, gas and taxes. However this is the only path that gives us access to gain truth and empathy.
Nothing is random, nothing happens by accident in the Universe. We have to be willing to change because without accessing consciousness, transmutation of the body is not possible.
As more of us are accessing consciousness, soon we will be thinking in multidimensional perceptions... As the vibration of the Earth and of our bodies increase frequency, matter will change accordingly. The Black Archons will not have power over us any more. In a period of acute social and political upheaval as is today, the Archons are afraid of us now. In this time of change it is best to have an open heart, let go of attachments, surrender and allow, because each solar storm is opening a path to higher vibrational frequencies.

"The most important characteristic for the survival of the species is not strength or intelligence but adaptability to change."   ~~Darwin~~

There is a selection of the mature souls that will move up to the 4th and 5th dimension.

Clairsentience, Clairvoyance and Clairaudience are still coming on line. These psychic capabilities were considered normal in the ancient world. Humanity must learn to rely upon and trust its own "magnetometer" as the information is being given by the higher self to the soul. Learn to listen to the soul.

True SOURCE LIGHT functions without equipment. Every week now, a new powerful flow of positive radiation from SOURCE LIGHT is coming to us and to the Earth, to bring access to higher states. Transformative experiences are happening almost daily physically, mentally and emotionally. Shifting is occurring to all aspects of creation.

Our beautiful and unique planet has such a variety of animals and plant life and different species of humans and skin color. Everything on Earth is being affected.

Our planet itself carries out "self cleansing" in an accelerated mode.
Mother Earth is also cleansing through unceasing tectonic activities, for example earthquakes.
On May/15/22 there were 78 earthquakes within 24 hours of 1.5 or greater on the Richter scale.
One in Sakhalin, Russia was 5.4, and one in Bengkulu in Indonesia was at 5.8.
On February 6, 2023 the Earth experienced higher extreme turbulence. There were 135 earthquakes within 24 hours and many have been at very high magnitude: "6.0", "6.7", "7.5", and "7.8". Solar winds are above normal.

Intensive hurricanes, powerful tornadoes, floods, united with our political turmoil, social discord etc., are all circumstances of cleansing.
Many thousands of years of "filth" are being exposed and are being cleaned up. We must be calm and focused and continue to develop a higher consciousness. In the silence of our inner world, we can begin to hear our soul and higher self speak to us as we reconnect to DIVINE SOURCE and DIVINE INTEGRITY!

I had learned from Father Quintin Walsh many years ago about the Akashic records.
These are the records of all our past lives and are stored in our heart Chakra united with our voice and DNA.
Our soul takes it when we leave the body, and our records are never lost.
We are the sacred incarnated vessel of the Source of all there is.

"There has never been a time when you have not existed, nor will there be a time when we will cease to exist. As the same person inhabits the body through childhood, youth, and old age, so too at time of death he attains another body. The wise are not deluded by these changes."

~~from the Bhagavad-Gita~~

Today July 11, 2022 our sun blasted two very strong solar flares during the night and in the morning. There have been 107 Earthquakes in the past 24 hours in the 3.0 to 4.0 and one of 5.0. On Wednesday July 13 we will have a "Super full Moon" and activations to our planet can become stronger during a full Moon. At that time Light activation experiences may escalate.

It is best not to engage in multitasking and hard work. We must engage in activities that are calming to our nervous system. The strong waves of Light cleanse and upgrade our DNA and cellular records continually.

———◦⊰⊹⊱◦———

I have known for many years that we are multidimensional beings, but I had never grasped the magnitude of what that really was, until I read the article below, and my mind opened up to the immensity of the soul!

I found this from an article by the Pleiadians, in 2019 and I like it so much that I read it often. It makes me realize how big we are and how complex is GOD Creation.

"States of Purification"

Friends of Earth!

Indeed we are with you in power and strength!

It is important to understand and Grasp the Enormity of the soul essence in your being!
The soul has lived lives simultaneously in aspects that are not aware of one another, yet can feel the other in vibrational frequency.

The wisdom of one aspect may be transmuted to another part that needs this information now. This is another method of intuition.

As each story ends, the soul aspects will integrate into the next vibrational frequency that matches the story completely.

Each soul now has fully integrated with you, or it is close to completion, as this is the cycle of the ages.
You that remain in form are the Gifted Ones!

Regardless of how little you believe yourself to know and understand, you are the highest vibration of your own soul!

By focusing your attention on the maintenance of your form in a state of purification, you will maximize your chances of completing the process of interconnectedness with the integration of consciousness which is, in essence, you in other densities.

Now you are ready! You have completed the purpose of arrival in this dream.

Your Chakra will open and the Akashic records will combine as the stories you have lived since Creation.

These energy points will guide you as you focus on intentions and stay present!
Indeed you are present to Ascend and many that you judge will become your beloved family and friends that arrived on Planet Earth with you to mirror roles for one another!
These roles have gifted you the opportunity to expand in consciousness and move forward into more Light!

You are the infinite light that you will rediscover and remember!
Indeed you are LOVE and so LOVED!

We meet you soon in raising!
Love one another and seek peace!

———◄►◄◊►►———

"It is paradoxical, yet true to say that the more we know, the more ignorant we become, in the absolute sense, for this is only through enlightenment that we become conscious of our limitations. Precisely one of the most gratifying results of intellectual evolution is the continuous opening up of new grater prospects."

~~Nicola Tesla~~

Many people on our planet do not believe that there are other different species of beings in the Universe.

When I look up to the sky from a place away from the city lights, the sky is full of stars impossible to be counted, I know that the universe is full of life. Each star has to have more than one planet where the beginning of all types of life is developing or thriving.

It is impossible for me to believe that Planet Earth is the only planet where many types of life form live (the way I was taught in school). I am actually convinced that there are stars that are much larger and older than our Sun with planets full of beings much more advanced than us.

The Khazarian cabal has erased all the knowledge we had about our encounters with extraterrestrials and our history, interacting with those benevolent beings.
We have been genetically combined with ET species.

The incredible mismanaged dispersal of wealth and human enslavement, needs to be disclosed otherwise it will never end.

People all over the Earth need to know there has been a war waged for millions of years in this Universe. It has been for total control of the Earth between the Anunnaki, Draconians, Syrians,

Grays, and unseen malevolent entities known as Archons, fighting against several benevolent races including those from the Andromeda Galaxy (which is bigger than the Milky Way Galaxy), the Pleiadians, and Arcturians, also the Galactic Federation and many others. They come from other Universes and are here incarnated and not, they came to help us and the Earth, to end this war.

There are many players in this monumental undertaking for the Liberation of Earth.
The negative ET races who went against the Law of the Universe are being recycled back to the Source, some are being rehabilitated, others have been recalled, and others are being sent back to their home planets.

The Light Alliance is in control and watching and waiting with excitement and optimism for humanity to completely liberate planet Earth from the dominion of the Khazarian Mafia/Archons.

The Event will happen, it is written in the Book of Revelations, and it is in many biblical scriptures, and ancient texts.

Below is a little piece from the EMERALD TABLETS written around 36,000 B.C. by Thoth the Atlantean, Priest-King. This manuscript precedes any Egyptian writings.

WE ARE THE CHILDREN OF THE INFINITE COSMIC LIGHT!

"We are truly Light, Sun of the Great Sun.

When we reach wisdom, we become truly aware of our communion with the Light.
One must rise above the darkness and become one with the Light and one with the Stars.
Always follow the path of wisdom; only in this way will we be able to rise from below.
Only with order will you be one with the whole.
Order and balance are the law of the Cosmos. Follow them and you will be one with the whole…
You are the spark of the Flame."

Hail to the man who went through life always helping others, knowing no fear, and to whom aggressiveness and resentment are alien. -
~Albert Einstein, (14 Mar 1879-1955)~

I recite many prayers, affirmations and Mantra that I repeat daily. This one is my favorite and I have been repeating it every morning for more than 4 years now. It gives me power and strength.

"In this now, right this second, in this very moment, I choose to heal all fear, trauma and limitation standing in the way of fulfilling my highest potential and purpose in this lifetime.

I ask my higher self and I AM presence, in co-creation with the elemental Angelic, Ascended cosmic and ancestral realms to assist me in healing transmuting, and transcending all limitation, separation and amnesia preventing me from embodying my full integrated sacred masculine and feminine power, love and wisdom.

In this now, right this Second in this very moment, I ask that my higher self and I AM presence direct my healing such that I embody and offer the optimal vibration in any circumstance; that I manifest optimal health and wholeness in and through my physical, emotional, mental and etheric bodies.

That I fluidly evolve according to my path and purpose; that I awaken and regain all my memories and gifts held in my casual body; that I function with energy, clarity, balance, effectiveness, and efficiency in the changing energy of Mother Earth.

That I integrate the healing, its lessons, and its energy in the most beneficial way possible; and that all of this unfolds at the most conductive yet expeditious pace possible.

In this now, right this second, in this very moment, I intend that in healing myself in this moment I heal other aspects of other consciousness trapped in time who, through the domino effect, will attain the freedom to return to where they belong, to heal and to contribute to the unity we seek to co-create in order to graduate from this plane and take our next evolutionary steps.

In the name of my I AM that I AM in this now. In the name of my Soul Presence, in this NOW in the name of the Light Forces in this NOW, I cancel and nullify all agreements and contracts with the Dark Forces in this NOW. All these agreements and contracts are null and void, regardless of their content, consequences and my subconscious programming.

In this NOW, I now release all belief systems that I no longer need that do not serve my higher purpose. In this NOW in this exact moment in the ever present co-creating moment, with my free will, I now declare myself free from all influences of the dark forces, now and forever, in this NOW at this exact moment, in the ever-present
co-creating moment.

I now decree and command full conscious cooperation between myself and the LIGHT FORCES. In this NOW, at this exact moment, in the ever-present co-creating moment.
I now decree and command that my life is guided in full alignment with the Divine Plan in this NOW.

I now decree and command that miracles manifest in my life in a way that will manifest happiness for me and everyone involved in this NOW.

Right this second in this very moment, in this NOW I offer healing, compassion and forgiveness wherever it is needed or wanted to free all aspects of consciousness including my own, to allow all to return to its natural order in wholeness and unity! And so it is.

"Concerning matter, we have been all wrong. What we have called matter is energy, whose vibration has been so lowered as to be perceptible to the senses. There is no matter."

~~Albert Einstein~~

We have so much light in us that we could light up a big city for many weeks.

Our Aura is a rainbow of colors that extends many feet away from our physical body. The Etheric body "Ether" is the state between energy and matter.

Each layer of the Aura has a location, color, brightness, form, density, fluidity and function. There are seven layers and each layer is associated with one of the major chakras in our body.

Every aspect of our bodies is being redesigned by the LIGHT, yet we must be willing to receive and acclimate to it.

LIGHT comes, we receive it, we adjust to it.
We are being prepared for a much more profound cosmic event to occur. Maybe it will be something we never could have imagined.

We have the power to change our planetary experiences, individually and collectively.
As a much larger amount of humanity awakens to Divine Principles, we will be closer to the "Golden Age."
We are blessed by Cosmic LIGHT!
Mankind is at a crossroad.
Let's connect to the GREAT SPIRIT/INFINITE PRESENCE/SOURCE! We are one with everything.

"Listen, I tell you a mystery: We will not all sleep, but we will all be changed—in a flash, in the twinkling of an eye, at the last trumpet. For the trumpet will sound, the dead will rise imperishable, and we will be changed."

~~1 Corinthians

Breaking News Of 04:00 PM CET, 25 July 2022

Official statement conveyed by Archangel Michael and Guan Yin, the new World Mother, the Karmic Council's Head, for the information and public disclosure:

The systemic evolutionary bifurcation point, as a result of which the equation of the Absolute will be self-solved, and the only collective force that will become the dominant and controlling force on earth, is scheduled for December 22, 2022, at 00:48 am.

The format of passing the bifurcation point would be a karmic duel between the two power poles of the modern world. From a technical point of view, it will be a spiritual-energy confrontation, where, as in the balance, the pole that is stronger in all respects will be determined.

According to the karmic law, this duel will be conducted exclusively according to the cosmic rules. The violator will be credited with a technical defeat.

The passage of the Great Bifurcation Point is an event of the Subtle World, but its physical component is also present. Based on the results of self-organization of the Bifurcation Point, the Equation of the Absolute in the 3D planetary format will be solved. That is, the identification of the only Dominant Force and Vector that will implement the evolutionary programs of Gaia and the Absolute on the Renewed Earth.

Soon after 22/12/22, the results will become noticeable on the physical plane, especially in the outcome of the hybrid Third World War. Its termination can be predicted in the first half of 2023. And it will be a direct consequence of the result of the karmic-evolutionary Duel of the World Poles.

⸺◦⦂⟨⬦⟩⦂◦⸺

Until now we have lived in a Babylonian slave system, our lives have been very hard.
As we approached the duel date, I prayed for all to be well with the forces of Light and I continued to recite Mantras for Peace on Earth.

All that was done in the dark eventually is coming to light. Never before have we been irradiated every day with such incredible high frequencies of Quantum Light from the Galactic center. Spirituality and consciousness. As the benevolent beings are coming into close contact within us, tuning into each other and affecting the 3rd and 4th dimensional consciousness of earthlings. The errors of the system, of deceit, betrayal, meanness, addiction, crime, racial, national and religious enmity, make us see even more the ugliness and filth that have been disseminated in our world, now our Earth and us needs to be cured.

On December 17, 2022 at 20:05 CET through a single hierarchical channel the results of a karmic duel of the two worlds was announced:

The duel between the forces of Light and the forces of Evil which was scheduled for December 22 took place instead in the night between December 15 and 16, because the forces of Evil without an advance notice struck a tremendous blow against the forces of the Light on December 15th at 09:01 PM. and the battle lasted until 04:00 AM. on 16 December. The assault was carried out on Black Shasta mountain in California.

The forces of Light were able to counterattack very quickly.
According to Lev, the confrontation was so devastating that the Dark Pole on the Subtle Plane of the Earth exploded into pieces.
The Pole of the Light was also damaged but remained fully viable. The strength and power of the LIGHT was disproportionately superior to the Dark one, the wicked had no chance of victory.
The victory of the Forces of Light in this karmic duel cleared a lot of space around the Earth, before the Winter Solstice, the wonderful result passed the bifurcation point.

"The resistance of the dark forces and the consequences of their actions will affect us for some time, but they will no longer be so destructive. From now on the Earth will have only one Pole of LIGHT, which means that the presence of the Pole of Darkness finally ends.
All efforts of the Co-creators and Hierarchs of the Upper Light will now be focused on clearing the Earth's of Hell in an accelerated manner."
Darks and their elites in the still-functioning Power Pyramid, did not recognize and did not accept defeat in the Duel of Worlds, but their stronghold is losing strength every day more.

The LIGHT Forces do not relax, as there is still a lot of hard work ahead.
They are more actively rescuing humanity, filling people and Earth with higher frequencies energies for easing their move to 4D/5D.
Those of us, whose consciousness, thanks to intensive inner work, are ready to receive the higher energies of 4D and 5D's quantum fields, grow Spiritually much faster. We are filling ourselves with new knowledge about the Earth's Transition to the Fifth dimension.

The Pleiadians will have a major role in the Earth's restoration process. The so long awaited new "Golden Age" or "Sattva Yuga" is getting closer.

—◇◦◦◦◇◦◦◇◦◇◦—

In January 22, 2022 I found this writing from the Pleiadians

Friends of Great Light!"
Everything is changing.
It" always has, as change is the key to your transformation into the higher realms.
As you once stood in Majesty and observed the lower frequencies, you understood that the change would be necessary for growth and understanding of the humans that seemed to find such chaos.

You decided to become one to fully grasp the experience. In moments now, you cannot comprehend this. But you will, beloved ones, you will indeed.

The transformation taking place within you is reflected in the Collective.
This is by Divine design as you remain to be holographic in nature and creation, and human interaction will begin to crumble unilaterally as the wake of the Grand Shift of humanity splits your reality to the greatest change yet.

You have created the phrase collectively that "the best is yet to come." Indeed, this will be a most welcomed change for all beings. Humans everywhere are experiencing the process of higher levels of awareness as the filters are blown away with the endless lies created by those in control in each waking dream.

You call out for the truth. You have been answered. Are you listening? Are you watching? For those searching for a lifeboat of doctrine or any imaginary rope to pull you into a space where you are judged on the very lies made to control you, sit down and breathe. You will be fine.

All that is not true will not remain. What will remain beloved ones is truth and knowledge of your true state of being in origin. You are not human. You are a Soul who has arrived with a great purpose. Only you can discover this purpose. If you arrive to experience blindness and lack, then flow with this. Nothing can happen without a sacred reason.

Nothing is by accident. Have you considered that your soul chooses the many obstacles you perceive, as a method to awaken you? You are fully able to align with your soul and find the wisdom of the higher realms.

You are changing beloved ones! And the feeling of unrest that is stirring through the Collective Consciousness will begin to play out as a catalyst to remove any obstacles that restrict your personal freedom! Come and fly again!

We love you so!

On March 21, 22, 23-- 2023 the Co-Creators will install in the Earth's Logos, a new permanent Matrix with 4th and 5th Dimensional software…

Well it didn't go too well with the installation of the new permanent 5D Matrix. While the co-creators were installing it, there was too much evil Karma still accumulated on the 3D Earth and also the Karma created by the wars and other daily evil. So the old Matrix crashed with the new Matrix. The co-creators had to immediately stop the download of the new Matrix or the Earth and all of us would have perished.

According to Lev: The deposits of karma in us are too large because we accumulated it over many incarnations and continue to replenish every minute by our emotions, thoughts, and unwillingness to transform. The worst thing is that we deliberately strangle and kill those who seek to change even our kids and others nearest and dearest.

The only way left for co-Creators to save us from ourselves is to let the high frequencies quantum streams do their work on us. Squeeze out and extract all the negativity from our physical and Subtle Bodies, from the most deepest hidden depths of our Soul, and present it to us as a choice: either we well clean ourselves from this toxic taint, or together with it we will be cleansed by high-frequencies quantum flows.
Powerful cosmic radiation of quantum Light works like mighty distillation, squeezing out of us and the entire civilization, everything that leaves no chance to survive in 4D and 5D and in this regard April 2023 will be no exception.

So through the entire month of April we went through so many changes with days of summer heat and other days of cold. The quantum waves definitely squeezed out all that Karma, and for some of us it was unbearable.
Here is the link where to read the rest:
https://www.disclosurenews.it/malaise-the-great-quantum-transition-lev/

April 28- 2023 Argorians update From Lev

While Co-Creators are working on the new 5D Earth's Logos (see <u>Ram</u>, DNI, April 27, 2023), friendly space races continue 24/7 gradually dismantle our three-dimensional world at the quantum and atomic-molecular level and simultaneously expand the 5D space on our planet's Subtle Plane. The Argorians' space fleet and two Siriusians' tracking stations on the Moon (one at 23D and other is its neutrino counterpart) daily report their contribution and current developments.

On 28 April at 09:38 AM CET, a neutrino tracking station signaled that new powerful high-frequency streams from Gerios Galaxy entered Earth. They ruptured some fragments of the cosmic header and lateral system, despite protective measures. Two rescue teams of Argorians immediately began to restore the breached parts.

At the same time, they warned that the frequencies will increase, and the quantum flows will become even denser and increase the strength of Solar storms. Unfortunately, on Earth, monitoring data is often deliberately underestimated and falsified in order not to create unnecessary problems for their electronic System.
The maximum expansion of the planet's Quasitron (a single system of Portals in the Bermuda Triangle, the Mediterranean and the Devil's Sea) increases the power exchange between dimensions, and, respectively, all energy channels and chakras are actively expanding in the human body.

Intense saturation of the Earth's space with high vibrations adversely affects our general physical and mental state, causing painful reactions in the form of fever or chills, internal spasms and exacerbation of chronic diseases. Gradually, all this will go away, the cells are constantly being cleaned and upgraded. It takes time to transform and adapt to new energies that were not previously on the planet….

We notice how time has accelerated; events are flashing like a kaleidoscope, and how tough it is to orient yourself in what is happening. Our feelings, attitude and consciousness are changing. The properties of the physical corpus and Subtle Bodies expand. Old information is being actively withdrawn from each cell, and it is increasingly difficult for us to recall past events. The freed space is gradually and carefully filled with new data, updating those who consciously perceive quantum restructuring and are ready for it.

The compositions of water, air, and minerals are changing. This causes responses in all living organisms, forcing either adaptation to 5D environmental parameters, or to the heavy diseases and premature death.

If our civilization lived in harmony with the Source, everything would be easier and softer. The spiritless, soulless and techno geek path of development led to the restoration of late Atlantis, whose Black elite continues to incarnate from the astral and etheric planes. Under their leadership, scientists began to interfere with the genome of Nature, disrupting the flow of cosmic energies, changing the properties of plants and animals, and also got into human DNA. But Co-Creators will remove all distortions. Anything that does not comply with the space Golden Standards and Pleroma's codes will leave the planet. Only Light carriers will be able to live in the fifth dimension. We are on the verge of active changes, the process has launched, the new life support system is on, and it's the 3D System final countdown.
https://www.disclosurenews.it/argorians-update-28-april-2023-lev/

September 10- 2023 By Lev

Pray that large percentages of humanity choose to embrace and acclamate to the LIGHT of SPIRITUAL ATTUNEMENT.

In "Kali Yuga" only ¼ of humanity choose to embrace and acclamate to the LIGHT of SPIRITUAL ATTUNEMENT.
Most others do not care and do not even understand the concept of HIGHER KNOWLEDGE".

As people around the world who are not directly affected by these events, they continue to shop, attend sporting events, watch popular movies in theaters, and generally do not care about what has occurred in these countries and are just glad that it did not happen where they Live.
Here is a reminder of what was told to humanity thousands of years ago by Yeshua Ha Messiah ("What most of mankind claims to respect as a "World Savior " and"Prophet").

One of the many prophetic teachings he gave was: "At the end of the age, there will be earthquakes in rare places".

We are at the end of "Kali Yuga" ("Age of Chaos") and at the point of what is known as the "Twilight of Kali Yuga" when planetary conditions are at their most critical point, prior to the beginning of the next age.
Massive earthquakes could occur in any place! When magnitudes reach "6.0", they are considered very powerful, and of course, over "6.0" and above can be tremendously destructive.

At the same time that Earth is thrashing as it is being cleansed of its cellular records—its atomic particles of ages of disharmony—it is being sent enormous doses of SOURCE LIGHT for NEW KNOWLEDGE and HIGHER COLLECTIVE CONSCIOUSNESS to manifest.

Yet, as is being constantly stated, humanity is at a "Choice Point". Higher percentages of mankind can elevate, or there can be continual devolution into total destruction.

I shall reiterate a verse from the BHAGAVAD GITA (which Yeshua Ha Messiah was very familiar with due to his travels and studies in Bharata ["India"]) which states:

"Humanity has been accorded free will and thus may adopt whatever channel—good or bad—through which they want their lives to flow."
He also said "The attainment of a spiritual consciousness is more important than material gain and prestigious titles."

The quakes that have ravaged Syria, Turkey, Morocco, and now that are beginning to "knock on the door" of the USA, can happen at any time in any place.
Remember the horrific quake and tsunami in Bharata (India) in 2004 where hundreds of thousands of people died.

This message is not to cause anyone fear and panic. It is meant to elevate consciousness and to focus attention more avidly on the SPIRITUAL FREQUENCY of SOURCE while we engage in regular daily activities as well.

After all, relaxation from stress and strain is vital in maintaining physical, mental, and emotional wellness.
However, we must also observe what is happening in the world because humanity is being given lots of "wake-up calls".

Earth is not just merely being "dusted" and "polished".
It is being completed "re-constructed", and every aspect of creation upon "Her" is being transformed.
disclosurenews.it/cosmic-frequency-news-9-september-2023-earth-in-chaos/

Cosmic Frequency News 16 November 2023 – Chaos Above And Below! – By Dr. Schavi M. Ali

https://www.disclosurenews.it/cosmic-frequency-news-16-november-2023-chaos-above-and-below/

Several filaments on our Sun are poised for eruptions with "M-Class" solar flares that are Earth-directed, and the planetary magnetosphere is completely enveloped in dense particle plasma with lots of it building on the far side of the magnetosphere indicating that it is emerging from outside of our solar system.

There is, thus, a heaviness in today's atmosphere after the fiery energy of yesterday's Tropical Sagittarius Moon which had many people having a very active day with lots of programs and projects to work on.
There may be more exhaustion, aches and pains, and other symptoms today of the cosmic energetics. Mars still in Scorpio is in conjunction with our Sun in Scorpio—one of whose warnings deals with anger, issues of justice, and increased warfare…

We need to learn how to properly "ground" as situations continue to occur on our Earth. When people are calm, they can think and act correctly and not be embroiled in chaos as their nervous systems are balanced.
As Sri Krishna taught: "A healthy nervous system equals a profound life force, and this eventuates into unlimited power which then becomes the realization of ETERNAL BEING while yet in the material body" (BHAGAVAD GITA).

Cosmic events are stirring our planet more profoundly as today's longer article mentions. However, this is what has occurred in terms of earthquakes so far this year:

112 in past 24 hours
954 in past 7 days
4,303 in past 30 days
51,880 in past 365 days

Biggest Quakes This Year:

Today so far: "5.7" in Indonesia
This week: "6.1" in South Indian Ocean
This month: "7.1" in Indonesia
This year: "7.8" (which some Seismologists say was actually "8.0") in Syria and Turkey

CFN Energy Supplement

"Light Activation Experiences ("LAE")—commonly known as "Ascension Symptoms"—come in waves. They thus arrive into the physical vessel, the emotions, and the mind in spirals, but each time, we are leveled up higher during the spiraling process. One day, for example, you may feel loaded with energy, and you are able to perform lots of tasks.

Perhaps you even slept wonderfully well the night before. Then, suddenly, the next day you may feel as if you have plummeted down, and you are exhausted, have no appetite, are anxious, notice a rapid heart rhythm, have a strange ache or pain in a muscle or joint, and you may have heat sensations and some sweating. There may be other signs of shifting energy as well.
It can be very frustrating to feel fantastic one day and lousy the next day—over and over again like the up and down movement of a roller coaster. However, the planet "Herself" is also shifting up and down—one day "She" is relatively calm in vibrational frequency and amplitude, and the next day there is strong stirring, pounding, and swirling. Humanity and the Earth are in synchronization.

We must transform with "Her", and the process—which I often refer to as "contractions" during our "re-birthing" experience—is enormous. Everyone is now experiencing some aspects of this dramatic change in human and planetary existence—some at more intense levels than others.

Many are, of course, relegating the "symptoms" to working too hard on a job or to the aging process or to an illness developing or to being influenced by world situations, etc. Few really understand and accept that there is a hard-to-believe amazing change happening in creation, and human beings are a major part of the change.

Many do not want to hear that more rest is required in order to acclimate to the changes that are occurring—just like a pregnant woman has to have more rest as her pregnancy moves along in months and as she may experience discomforts. We are in a new reality—a different "Now"—a space/time continuum that was always meant to happen according to how mankind chose to use its free will.

Thus, this is the time frame of (and here come those usual terms): CLEARING-OUT OLD PROGRAMMING, CLEANSING CELLULAR MEMORY, and CREATING NEW MULTI-STRANDED DNA IN EACH CELL AND THUS MAKING NEW CELLULAR RECORDS. As the months of November and December of 2023 move along, and as the next years traverse the cosmos, more and more of humanity will elevate. It is ordained to be so by SOURCE. Those who are veterans to the process because they have been having shifting experiences for many years can assist others to understand what is happening—if the "others" are open to the information.

HIGHER KNOWLEDGE will come only to those who engage SOURCE FREQUENCY—not to those who are wandering in a wilderness of ignorance and confusion. As Yeshua Ha Messiah said to those who have been dealing with lower vibrating consciousness from others: "COME OUT FROM AMONG THEM". Be more self-nurturing in these times. Be an "ashram" of PEACE.

CHAPTER 13

CHAOS ABOVE AND CHAOS BELOW

WHAT TO EXPECT NEXT?

August 29--2023 From Benjamin Fulford

>EMERGENCY BROADCAST SYSTEM<
>PENTAGON >EXPOSURE >CONNECTED
>TO >PLANDEMIC >WORLDWIDE< >LOCKDOWN >AGENDA 21 2023<

Be ready for the >EVENTS< unfold to >EXPOSE the >CABAL >DEEP STATE >AGENDAS >You will know we are right around the corner from a NEW WORLD. Do not get angry, and do not panic. This is NEEDED. Yes it TRULY must HAPPEN like this so we can TRANSITION to a BRIGHT FUTURE. It is part of the SCRIPT and LAST PHASE of, indeed gut wrenching, AWAKENING movie that was necessary to AWAKEN the masses.

This will ensure EVERYONE is SAFELY placed in their HOME and able to WITNESS the HISTORICAL moment that REVEALS all of the TRUTHS, cover ups etc. through the >E B S < which is imminent. All is SCHEDULED to HAPPEN. so get PREPARED. Circle the DATE on your calendar and PLEASE pay attention. There must be a TEST and then a review of all OCCURRENCES and ACTIVITIES. The possible implications on a NATIONAL and GLOBAL level can be quite COMPLICATED so things must be in ALIGNMENT to the PROTOCOLS.
Yes there are many consequences if things aren't done with PRECISION and perfection. This is the PRACTICE run before the REAL ONE to see responses and accuracy to what is FORTHCOMING which changes >HUMANITY< We hear the SCHEDULE is now FINALLY firm, but again I'm just the MESSENGER.

Be READY to adjust if needed in regards to possible TIME changes. Only a SELECT few know the MOMENT of exact and precise TIMING of EVENTS. For SECURITY and other obvious reasons it must be properly kept PRIVATE.
Again the >E B S< is going to AIR playing an 8 HOUR VIDEO. It will be replaying 3 TIMES a day for 10 DAYS, Communication DARKNESS.

During those 10 DAYS of Communication DARKNESS the following things will happen. We will RECEIVE 7 >TRUMPETS< aka >E B S< text MESSAGES on our PHONES alerting US to tune into our TV at this TIME. Our PHONES will only work for 911 and we are INFORMED the Signal App, which is MILITARY encrypted will be available. Our TV's will only show 3 EXPLANATORY MOVIES on a continuous loop for the 10 DAYS.
It will cover topics of ARRESTS, TRIBUNALS, FRAUD CORRUPTION, PEDOPHILIA etc Our INTERNET will not work during that TIME. Our ATMs will not work.
 After the 10 DAYS of Communication DARKNESS, we will connect to a new QUANTUM internet.
People are urged to STOCK up on at least THREE WEEKS of FOOD and WATER. We are PROMISED the new Star-link Internet System by the end of the month. Again I repeat be PREPARED with FOOD, WATER, TOILET PAPER, generators etc. for this >GREAT AWAKENING< REVEAL. As we speak the TEAMS coordinating this IMPORTANT

HISTORIC EVENT are revamping the >E.B.S< to ensure the utmost SECURITY for all INVOLVED so remain PATIENT as things get finalized.

They want to make CERTAIN there are not any interference of any sort at all. Those making the PLAN want no ONE to PANIC because it's simply the release of the TRUTH. After the >E B S < and we've gone through the 10 DAYS mainstream media BLACKOUT and sat through all the 24/7, [eight hours long movies] do we go back to NORMAL like business as usual?

Answer is: After >E B S< and the 8 hours long 24 7, movies all will change. The LIFE support, attached to the old and EVIL systems will be PULLED. >HUMANITY,< and >PLANET EARTH< simultaneously move to QUANTUM reality consciousness system [PEACE and PROSPERITY].

The END of FINANCIAL and HUMAN consciousness ENSLAVEMENT. Old systems of GOVERNMENT, EDUCATION, FINANCE, HEALTH, TRADE and COMMERCE etc., will all be DISMANTLED and REPLACED.

We will have a new CURRENCY called the USN US NOTE and GOLD backed.

The TIME is now to ALERT as many who will LISTEN. Do not have too much PRIDE. Go WARN those you LOVE even though they think you're CRAZY. Your GOAL for OTHERS is TRULY to HELP absorb the SHOCK of what is COMING.

Stay strong >PATRIOTS and >STOCK BACK UP ON >FOOD. >RESOURCES. >SUPPLIES for the >EVENTS< happening and coming. >We are inside the >STORM< MAJOR >EVENTS HITTING<

Everything is leading to >>>MILITARY<<<
   //>CRIMES >AGAINST >HUMANITY</   /

Q) October 9- 2023 ----------- Changes are happening in our world!

Q World operations were placed long ago and.]]] INFILTRATION [[[instead of invasion was the first stages of the PLAN<

>This involved inserting Military commanders into the Deep State regimens, just the same as [ds] ops placed by RINO'S

Republicans who were placed by the CABAL into Republican seats and so on." "" "

You think it's a coincidence General Q, (Charles Q. Brown jr.) Who was placed by Trump and appointed to be the Chairman of the Joint Chief of staff,... as the chairman he is the highest-ranking and most senior military officer in the United States Armed Forces<And now Charles Q. Brown Jr (who goes by the name CQ and commanders call him General Q) was also formally announced as President Joe Biden's nominee to

succeed General Mark Milley as the 21st chairman of the Joint Chiefs of Staff on May 25, 2023.

Why would Biden, the CIA, General Milley allow CQ to become the leader of the Pentagon and give him the highest Ranking in all U.S. Military Forces? Knowing he was placed by Trump?
What's happening?

Here is what's happening.
>TRUDEAU is being removed, his image is coming to light and government officials in India EXPOSED Trudeau as a coke head and moving drugs,.. and his airplane was seized for inspection (behind the scenes the Military Ops were tracking him and knew what was inside) _ Now lots of reports and direct info from Indian diplomats are telling the truth of Trudeau.
>CIA direct involvement in the COVID-19 cover-up and payments to Scientist to suppress The Gain of function (origins)
>Fauci [EXPOSURE] inside the CIA and pentagon numerous times, without signing in, nor going through security Protocols
and meeting with CIA officials to help coverup the COVID origins
>Impeachment of Biden begins. Major moves and EXPOSURE
>Hunter Biden indictments.
>
>General Mark Milley is behind removed and taking his place his is a Trump placed General Q_ (Charles Q. Brown) CQ
>Rupert Murdoch steps down from CNN world controlled News outlet
>Zelensky was denied resources and left Washington with nothing and is denied access to Congressional speech.
>Michael Bloomberg RESIGNs
>Daniel Andrews Victoria Premier steps down.
>Public release of documents showing the COVID-19 pandemic was a US Department of Defense Operation dating back to President Barack Obama.
>The EXPOSURE of the CIA happening in Russia, China, U S. On national levels enormous
>
>>>>>>
SO MUCH IS HAPPENING as across the world officials are stepping down
>
<.>Poland turns on UKRAINE and stops sending weapons and threatens to stop ask delivery of NATO resources through Poland
>African countries are fighting against CIA, M16 French controlled colonization of their countries and stealing of their resources
>Half of the world rejects the U.S. dollar

 i have been giving you updates of so many things happening that i can't write everything here now)

—

So why did Biden, Cia, Obama, [ds] Mil. Allow Trump to appoint General Q to lead the Pentagon and look over all over the U.S. military as Joint chief of staff?

>Well I have been telling you these White hats operations were taking place long ago and these military ops installed Trump and inside these operations TRUMP in return gave full (legal) power to the military to carry out legal Covert operations that included THE DEVOLUTION PLAN. CONTINUITY OF GOVERNMENT ////that's all going to lead to open military intervention inside the COLLAPSE>

TRUMP is giving you COMMS when he said General Milley should be Executed for Treason (this WARNING is for all the Deep State <)
>The Military Alliance placed TRUMP > TRUMP PLACED GENERAL Q INTO POSITION

_NOW CABAL> THE DEEP STATE CIA WORLD OPS IS IN MAJOR PANIC AND THEY ARE TRYING DESPERATELY TO CREATE A NUCLEAR WAR>>>

... BUT THEY WILL FAIL / but still we will come close to the NUCLEAR EVENTS I warned of / more over the coming planned pushed civil war by the democrats CIA Rockefellers regimen   / but everything will be EXPOSED in the end and the fall of U.S Military Industrial complex system COLLAPSE is coming and the Elites who control the war mongering system> BLACKROCK is going to collapse......

It's biblical and it's all connected to EPSTEIN. BIG TECH. ISRAEL , CCP. CI,. Bush's, Obama. HUMAN TRAFFICKING
World banks ECT etc.

Q_THE PLAN TO SAVE THE WORLD

———◦⊰◈◈⊱◦———

In 2021 I was waiting for the right time to add "Operation Compression" to my book. Unfortunately we were all immersed in the pandemic which did not allow us to think about anything else, because in those days people were witnessing all those deaths and how to stay alive..
Too many humans were dying in hospitals, or after having shots for Covid 19 and then the following shots that various doctors and governments forced humanity to take, if they wanted to continue to even have a job.
Now that many have woken up to the harsh reality of the facts, that these "vaccines" were and are weapons of mass destruction, perhaps now someone will believe that the universe has other intentions for Humans, and the LIGHT of the SOURCE GOD is changing us and awakening. The earth is not simply "dusted," it is "rebuilt," and every aspect of creation about "her" is transformed.

OPERATION COMPRESSION By Lev on 26 April, 2021

disclosure news.it

Immediately after Armageddon, on April 23, 2021, Galactic Committee
with 25D Siriusians and their space fleet in Earth orbit began Operation Compression.

Using energy reflectors on spacecraft, they compress the Earth's magnetic field to
accelerate the merging of 3D, 4D, and 5D into a single space (eon).
We should well understand that the success of Operation Compression and other
Galacom's ops, like the victory of the Light Forces at Armageddon, do not mean the
liberators in the streets, not the winners' flags on the balconies, not the bouquets,
kisses, and hugs.
For now, it's ONLY dismantling 3D debris, and terrain clearing for building a new planet.
The most important construction site is our Souls, thoughts, emotions, and actions. On
it, the main victories for many are yet ahead.

What's in store for us in 4D?
People, who have reached the level of frequencies of their bodies up to the fourth
dimension, are destined by the Galactic Committee to a better life.
Soon, they will be given back their Divine qualities. And the very first of them is the
possibility and right to live in the new reality in a young and healthy physical body.

How will it happen?
The fourth dimension is buffer space, intermediate between the third and fifth
dimensions.
3D is a world where consciousness is placed in dense matter. In 5D it is placed in the
Subtle matter. And between them is the space of neutral 4D, equally loyal to both dense
and Subtle matter.

The fourth dimension is used by the Higher Light Hierarchy to merge the two worlds. It
is the space where beings from neighboring densities meet.
Beings that evolve from the dense space into the Subtle Plane ascend into it from the
lower levels. They decompress their physical bodies and raise their vibrations.
From the higher dimensions, fellow Subtle beings descend into 4D. They have to
consciously lower their frequencies and condense their body.

The dense ones are dissolved, the Subtle ones, on the contrary, are condensed. Thus,
a platform is created where lower and higher beings get an opportunity to see each
other and establish contacts, cooperate and interact for the benefit of the evolution of
lower consciousness into higher levels.
How will we move into this space?
First, by expanding our consciousness to the fifth (astral) dimension at least.
Second, the physical body must vibrate at frequencies close to the fourth (etheric)
dimension, at least two-thirds of the 4D.

On the lower floors of the fourth dimension, we go through the procedure of matter code replacement. Our body is constantly being bombarded, like by hot steam, by high-frequency quantum waves which we need to get used to.
Dormant DNA strands in us are activated.
The frequency of our cells' vibration is increasing. They are infused with Light and crystallize it within themselves. This is how we are prepared to meet the Light.

Once our vibes reach a density of 4.5 to 4.6D, we will begin to make contact with entities of the 5th dimension on the site of the fourth dimension.
They are our highly evolved ancestors who have been waiting to be reunited with us as members of their family for many thousands of years.
Contact with them will be made by people (and groups of many of us) that have reached a level of frequency that will allow them to see higher counterparts, and feel comfortable in the presence of their radiance.
Although Higher Light Beings can lower their vibrations to the 3D level, they do not like to do so.
Our space is very messy for them right now. It's not good for their bodies. And that's why relatives are waiting for us halfway between our densities – in the fourth dimension.
What's waiting for us in 4D?
The most important thing is to be able to enjoy life. The ancestors waiting for us have the technology to dramatically raise the quality of our lives, to help us become spiritually and materially wealthy, and fulfilled.

Life in the higher dimensions bears little resemblance to what we live in now, and what the Archons and Dark Forces have accustomed us to.
There is no fear and division but trust and Love. There is unlimited freedom to do what the soul craves, what expands its boundaries and brings joy.

There is an opportunity to manifest ourselves in a society interested in our evolution, in the unfolding of all our Divine qualities.
The reality of the fourth-density is very fluid. We will have to get used to being able to create almost instantly whatever we want and need.
It will take us some time to fully understand it and to become capable of doing it.
Soon, the main attraction for us will be to learn how to apply this possibility more deeply, practically. We will start experimenting with space and our bodies.

At first, we will do it under the guidance and with help of our Curators. It is one of the many things they will teach us with great eagerness when they determined that we are ready.
All beings are in the fourth dimension only temporarily, for adaptation and training to live in more Subtle Planes: the fifth, sixth, seventh, and higher.
The higher dimension, the higher is the quality (taste) of life. People usually do not aim to stay in the training 4D for long. They try in every possible way to raise their body vibration to the 5D level.
However, even being in the fourth dimension is incredibly life-changing.

Who will we see first in 4D?

In this space, the long-awaited meeting of two origin civilizations of the Earth will take place. One is our Fifth Race that is now ascending to 4D and 5D through body transformation. The second is the civilization of our highly evolved relatives, such as the modern Lemurians.

They crossed into 5D several millennia ago but deliberately keep their physicality at the 4D level so they can be the first to meet us in the buffer zone, and teach everything we need to know.

On Earth, one of the strong Power Places is Mount Shasta. In 3D it is a dormant volcano 4,317 m high, one of the peaks of the Cascade Mountains located in California, USA.

The Indians believed in the special sacred properties of Shasta that gave them wisdom and physical health. Their shamans made ascents to the mountain to communicate with spirits, and for cleansing ceremonies.

Nowadays there are numerous pieces of evidence of the anomalous phenomena occurring on Shasta. Eye witnesses tell of tall humanoids in white robes emerging from under the ground.

People see strange lights and hear unusual sounds, songs, and melodies coming from the mountain. And also observe many UFOs landing or launching from the top of Shasta.

This is not a speculation, this is a reality. Shasta is one of the key sources of our planet's power. At the level of the fifth dimension, it is an embodied aspect of the Central Sun of our Local Universe.

It is a center for Spiritual Guides and Teachers from the Realm of Light. In the 5D of the mountain is Agartha which includes 120 underground Cities of Light.

It is also a haven for the survivors from ancient Lemuria, our kinsmen. They have long been, and now especially actively prepared to meet us, their friends and relatives, on the territory of 4D and to unite our two worlds into one.

As for the "tall humanoids", the original height of Lemurians before the cataclysms was about 3.6 meters. When their continent sank, the remaining humans were lifted into the fourth dimension, and their height was reduced to 2.1 to 2.4 meters.

Such they remain to this day. Although Lemurians evolved into the 5th dimension over time, they deliberately left enough physicality for the soon-to-be merger of the two civilizations.

Eyewitnesses were not wrong about UFOs either. Mount Shasta is not only home to Lemurians but also is an interplanetary, intergalactic and multi-dimensional Portal for many friendly cosmic races.

Today, Lemurians have a huge fleet of spaceships called the Silver Fleet. They use it to travel within the 5th dimension and higher in the Greater Space.

Spaceships are physical but can easily become invisible and silent in time of need to avoid being seen by ground armed forces. Ships are capable of changing their energy fields from the third to the fourth and fifth dimensions and back again if necessary.

Above Mount Shasta, there is also a huge City of Light called the Crystal City of the Seven Rays. It is an awesome creation that will be the first of the Cities of Light to descend into the physical world and be manifested on the surface of the Earth. However, we must be equal to their vibration. For soon, our two civilizations will reunite to build the long-awaited world of abundance, beauty, and perfection.

Under the leadership of the more advanced Light races, a collaborative effort will begin to recreate the once lost grand design of the Universe on planet Earth, only now on a permanent, eternal basis of the supremacy of Light.
Once, Lemuria possessed unique technologies which the current Lemurians have carefully preserved, adapted to modern conditions, and are ready to demonstrate to the new humanity.
To do the same are ready Siriusians, Pleiadians, Arcturians, and many others friendly races.
Most technology is, as before, controlled by thought power. In particular, know-how that makes 4D inhabitants young and healthy again.

In the higher dimensions, everyone looks 30 to 35 years old. Those with severely worn-out bodies will go through a rejuvenating procedure two or three times, gradually youthify the body to the desired level.
There will be some more nice bonuses of the fourth dimension (briefly).
New houses will be built for the residents of 4D with perfect crystal technology. The houses will be heated by small devices, the size of a mobile phone. One device can supply light and heat to an entire block.

People who have raised their frequencies to the fourth dimension level will no longer need our electricity and gas networks. Owners of country houses will not need to buy expensive fuel to maintain a comfortable temperature in the house. There will be as much light as is needed.
There will be a new, quantum Internet and telecommunications system. Computers will run on amino acid technology.

The know-how that will allow growing agricultural products without chemicals and fertilizers will be transferred, only by the energy of thought and emotions. The technologies of long-term storage of fresh harvest without freezing and canning will be provided. It will give huge cost savings.

There is a lot to tell… But we must understand that for living in 4D, we must raise our frequency and consciousness to this level.

To meet in 4D the members of our highly developed family, it is necessary to exercise daily the art of true Love. It begins with filling our hearts with love for selves, for others, and all beings. As the veil between 3D, 4D, and 5D continues to thin, a new world in all its glory will soon appear before those involved in the transition in a very physically tangible way.

But it requires us to be conscious, not to be distracted by negativity; concentrate on everything positive; raise our vibrations to the fourth dimension and be ready for the merger of the two worlds which can begin for us at any moment.

October 4, 2023

UPDATE!!! EBS & GESARA Countdown: Exposing Deep State & Banking Betrayal, QFS Revolution, 5D Earth Shift, Trump-Obama Power Play, Petrodollar's End, and Iraq's White Hat Alliance!
HIDDEN AGENDA

EBS, Martial Law, and Gesara: The Countdown Begins
The way things will unfold is meticulously planned. It starts with the EBS, the Emergency Broadcast System – the messenger of utmost importance. The Martial Law declaration will follow, all in line with the monumental GESARA announcement. Is the EBS preceding or succeeding the GESARA? Ah, the uncertainties remain, but one thing is certain: once Disclosures come to light, the Redemption Center Appointments will beckon.

The Redemption Centers: It's Not Just About the Money
Every single one of us will have to pass through these Redemption Centers. And it's not just for those fortunate to possess foreign currency. All the dark deeds, including military tribunals, public punishments, and much more, will finally be unveiled.

The Sinister Secrets of Banks Revealed
Our trusted financial institutions? The very banks that promised to safeguard our futures? They've betrayed us all. Not just you, but the generations before you. With the initiation of GESARA and NESARA, a financial reckoning is upon us. The mighty dollar will be returned to its rightful owners: the people

---

November 5th -- 2023 a post from Benjamin Fulford.

>AWAKENING<

The WAVE of Awakening and the HIGHER States of Consciousness continue to move All Across our Planet.

This is what the Dark ones Feared the most - - that the Population would AWAKEN - - and NO longer need them.

And their Biggest Fear is that NO one will want or need them any longer. That is the way it must be.

For as the Population of this planet Awakens more and more - - it is doing so as a result of All these Waves of ENERGY that have been coming in–Wave after Wave - - after Wave. With each continuing WAVE Becoming stronger and stronger. />>> >>> >>> /

And those on the Planet Being able to Raise their vibrations and receive these HIGHER Vibrational Frequencies coming in.

And in so doing - - the Awakening is Happening NOW. The Ascension process is in Full Swing.

And those of us - - the Lightworkers and Warriors - - We are the ones that are fully in the Ascension NOW - - in which we have heard as the first WAVE of Ascension is Already upon us.

We have not ascended fully yet - - but are in that Process NOW.

The SOLAR FLASH - - the Great EVENT - - will only happen in the DIVINE TIMING that recognizes the Consciousness of this Planet that has risen enough to BE able to handle this strong ENERGY coming in.

For if the ENERGY were to Happen NOW - - if the Solar Flash were to Happen NOW - - many across the planet would NOT Survive.

Even those of us - - many of us would likely NOT Survive It.

So it CANNOT Happen yet at this point.

But because of the continuing Waves of ENERGY that have been coming in - - We are moving closer and closer to Being able to Not only Withstand the Energies from the SOLAR FLASH - - but BE able to move Fully into our Ascension as a result of those Energies - - as HIGHER LIGHT BEINGS.

We have heard that Atlantis and Lemuria will rise again. But it will rise again within each and every one of us.

For us carry the Memories and the Remnants of those long ago in many ways forgotten Civilizations. But forgotten for NOT much longer.

For they are indeed Arising again. And we - - each and every one of us that listen to these words and resonate to these words: we are those Lemurians come back again.

BE able to utilize those tools that we had at that time. Just as many of us are becoming Aware of the Crystals again - - and the connection with the Consciousness within the Crystals.

And how Crystals can Hold Consciousness and Hold Remembrances. All of that is Returning.

Everything is exactly Happening in DIVINE TIMING and moving for the Full Resolution of the end of this old age of illusion and the Beginning of this NEW Sense of ONENESS.

No matter what happens - - We continue to BE who We Are.

And Everything will continue to take its course just as it is meant to. For WE All are in that DIVINE TIMING right NOW.

———◆◇◆◇◆———

November 26--2023 Disclosure continue From Q) Anon

CABLES>];
Inside the Military Cables and channels running there is a massive> WIRE.
"U.S. CORPORATION IS DONE. AND HAS FALLEN"

_This means all top major Countries are fully aware the United States corporation is bankrupt and the Biden administration is a fake administration and the CIA operations behind the scenes are fully collapsing and countries connected to the CIA regimen money laundering systems are under attack by white hats military alliance.

For the first time it's becoming clear to most of the politicians across the world, that The UNITED STATE'S COULD BE UNDER WHITE HATS MILITARY CONTROL AND OCCUPATION and the U.S. 2020 stolen/ rigged elections were/are actually U.S. military operations to EXPOSE a world wide corruption system that is connected to world banks, governments and their powerful Elites.

(Most ANONS. PATRIOT AND Q FOLLOWERS were already aware of this information years in advance.....The once conspiracy is now hitting the BEHIND THE SCENES world political heads and leaders and there is major PANIC....... That TRUMP intentionally got arrested and his indictments were all MILITARY OPERATIONS that will lead to EVIDENCE of Deep State military COUP and executive Order signed back in 2018 that connect to military occupation in cases of Future 2020 foreign elections interference protocols) /

_more over the PANIC hitting the EU is that EXPOSURE OF THE CIA INTERFERENCE (w/social Media/msm) IN ELECTIONS WILL ALSO BRING FORWARD EU COUNTRIES EXPOSING THEIR OWN STOLEN ELECTIONS AND COVER-UPS IN THE PAST12 YEARS and their corruption connected to WHO. CIA. NIH FAUCI. MSM LIES, PANDEMIC, VACCINES DEATHS.

BEHIND THE SCENES >] BILLIONAIRES ALLIANCE who lost Trillions in the past years to democratic movements, the PANDEMIC, THE MAN MADE VIRUS, CIA, COLOUR REVOLUTIONS ... Rigged EU financial reports and THREATS/ KILLINGS of their own families by the EU Deep State cabal regimens were never forgotten.

(Three years ago I told you that a billionaires ALLIANCE was growing and created by powerful Italian families that were working with the Trump Alliance and military Alliance and they were making sure that EXPOSURE of the PLANDEMIC would be revealed..

From vaccines to gain of function to governments connected to human trafficking, drug trade. Money laundering, War profiteering by EU Elites) _This billionaires' ALLIANCE is against WHO, BILL GATES, blocking out the sun agenda, the climate agenda and all things deep state .... And this ALLIANCE is bringing forward the GREAT AWAKENING MOVEMENT on a large scale.....

From military ops, to police investigation to lawsuit to SOCIAL MEDIA EXPOSURE>>>OPERATIONS INSIDE A.I. SECTORS that are going to attack the deep state A.I. agenda / EU Projects that protect the deep state. >> WEF/BILDERBERG GROUP/ SCHWAB the EU cia regimen in Switzerland   /

_
A massive STORM is COMING as almost every EU country including Australia. Canada and their military SETS are getting ready to arrest each other for TREASON..

Everyone is getting ready for World COUPS.
Arrest Wars.........And TRUMP/ MILITARY ALLIANCE holds the keys to the EXPOSURE> CIA> 2020 MILITARY COUP / EXPOSURE of the virus. > EPSTEIN<

_
Hello McAfee your terabytes of data and your hacking into world networks and [ds] government/ [ds] military intelligence computer systems was a work of art. a masterpiece admired by the USSF CHEYENNE MOUNTAINS+ALLIANCE

Q post #3432 November 7--2023 Q) The Storm Rider

This is not another 4 YEAR election....
"DRAIN THE SWAMP" does not simply refer to removal of those corrupt in DC....
GOD WINS.
Q
—
_This post is important! This post refers to deep state underground military bunkers .
Roads. Channels and networks that connect the world. From Africa to the middle east,
to the Vatican to Switzerland., Europe. Ukraine up to the north pole and back down to
Antarctica .

This is why disclosure of the DEEP DARK world operations and underground bunkers,
cities, biolabs and weapons, human trafficking networks need to be exposed.
And the EXPOSURE can only come from the collapse of the [DS] military and [DS]
intelligence that protect the Elites, CABAL and secret societies that run the world.

This is all>BIBLICAL and much farther into the past than most can even imagine and
fathom.

From the middle eastern ancient times of Iraq , Mesopotamia and Samarian Gods, to
Aliens, to fallen angels of the Bible, to Spirits, demons, dark blood lines ....
This all connects to evil Powers and corruption that ran cities and controlled lives and
civilizations.

Even the Khazars in 600s off the dark blood lines robbed and went from country to
country stealing identities and human trafficked and prayed to devil's in which
adrenochrome was consumed.
Eventually the khazars made it to Ukraine and stole the identity of the Ukrainians and
infiltrated the nation and later on the khazars did pledge allegiance to the Jewish people
and become followers of Judaism, but still pray to their Gods of death.
The khazars only wanted to take the Jewish power and slowly they infiltrated and stole
the identity of the Jews (these are some of the first military counterintelligence moves
made by the khazar military.. infiltrating societies) ...
Later on the Khazars would travel through Europe and leave their people to infiltrate
countries and nations. (history books say the khazars were defeated and disbanded but
that's only lies as today's KAZARIAN Mafia took control of book publishing industries
through Europe in 1700 1800s)
The Khazars /KHAZARIAN /KAZARIANs eventually moved across Europe and left their
family in Germany and these People became the Rothschilds of Frankfurt Germany....
Later on the KAZARIANs who are the Rothschilds, formed an alliance with JP Morgan
and created the U.S. federal reserve ([they] killed all their powerful rich oppositions on
a boat known as the TITANIC<.

The Rockefellers, who were 33 degree Mason Jesuits, came from the same close regions in Germany as the Rothschild's........ These powerful ELITES long gained power from the Vatican who was heavily infiltrated with the KAZARIANs, the Jesuits, mason's, knights of Malta military intelligence and much more..

(👆 I'm only giving you a small piece of the dark bloodlines and the true history of THE SWAMP <Elites)

_hidden beneath the Vatican is over 50 miles of books. technology, records of past civilization that have existed more than what fake history books are reporting..Nor what the HEAVILY REDACTED BIBLE had revealed.
(Imagine what CNN. The CIA. And governments. Military intelligence hides these days...

 NOW imagine how many times the bible was re-edited, remade over a thousand years with its first official production in 1455......Even back in the year 200 the pope and roman officials were arguing on how to hide information from the bible...
By the 1400s Jesus was pushed as Caucasian with blue eyes and blonde hair by the Popes and Vatican....

 In this time Cesare Borgia the son of the pope was rumored to be the lover of Leonardo da Vinci who painted Jesus in the image of Cesare Borgia, it's also known Borgia killed his brother for power and took his leadership. The real history of the Popes and power is connected to betrayal, killings, counter intelligence and feverish deception. ...There are so many mysterious disappearances that have taken place around the Pope's since the beginning.

*NOTE * Jesus was a real person with important super intelligence and his understanding of the true Spiritual energy that can bend reality and time.. Which means defeat death. TIME TRAVEL<   /
unfortunately a lot of his true teachings were hidden and suppressed<
Now there are over 20,000 different denominations of Christianity and they are all fighting about who is right and wrong.

>The Roman/Christian wars created the first banking Systems and military intelligence and massive human trafficking networks. They stole most of Europe and large parts of middle east gold, silver, arts resources and historical documents and his them under the Vatican vaults and caves...
To this day the Vatican museum and open vaults house over a trillion in arts and sculptures and jewelry and precious metals stones the public can see...But what you can't see is the hundreds of trillions of artifacts hidden in the Vatican underground caves, bunker's and vaults that stretch over 50 miles underground.

The Satanic power that infiltrated the Vatican and pedophile world networks is connected to KHAZARIAN powers, Mossad. Cia. Rockefellers and Rothschilds who helped MOSSAD CIA finance Robert Maxwell the father of Ghislaine Maxwell> and Jeffrey [EPSTEIN]

DRAIN THE SWAMP means draining the world of the SATANIC CABAL .

Q) the Storm Rider Official Page

]] INFILTRATION [[

The Military Alliance white hats ops had infiltrated the deep state projects and operations since the 70s
>That includes white hats placed inside the World Economic Forum
_CIA
_FBI
_DNC
_WORLD BANKS
_WORLD HEALTH ORGANIZATION
_NATO
_UNITED NATIONS
_THE VATICAN
_ CARNEGIE ENDOWMENT
_LOCKHEED MARTIN/SKUNK WORKS
_BOEING
_GENERAL DYNAMICS
_THE MILITARY INDUSTRIAL COMPLEX
_THE MILITARY CENSORSHIP COMPLEX SYSTEMS
_BIG TECH
_BIG PHARMA
_CDC
_WALL STREET
_WORLD ELITE FAMILIES AND REGIMENS
_MUSIC INDUSTRY
_THE WORLD SCIENTIFIC COMMUNITY

_NOW you will see white hat sleepers from country to country wake up and start to EXPOSE their industries, their governments, their Corporation, agencies and their own corrupt family's and regimens.

It's already happening and will grow.

From Congress going after Epstein to exposing Biden.
To Congress, Senate exposing the gain of function virus

To true insiders and CLASSIFIED agents in different sectors of military and INTELLIGENCE coming forward on corruption. Vaccine death exposure. To CIA, UFO, UAP retrieval Programs.

Even leaders who are heads of the climate change agenda will come forward and expose the hoax and corruption behind the UN / WEF climate change agenda.

Across the world in over a hundred countries, white hats sleepers are ACTIVATED<

They will all expose their government and regimens of corruption tied to vaccine deaths, cover ups. Fake pandemics.

From New Zealand to Europe to Mexico to the U.S. Canada and beyond the MASSIVE Great Awakening operations are taking place.
The next 4_7 months will be the fastest awakening movements to take place on earth.

Elon Musk took down Disney CEO who challenged him within a few hours by having tens of millions imminently boycotting Disney.
" Go fuck your self) Musk said... And also gave COMMS that the world will choose the right.
(He was actually giving COMMS letting everyone know he can maneuver an army of global Patriots against the deep state in a moments notice<)
_recently now AJ confirmed that Elon Musk was in fact working with USSF and TRUMP placed him into command (i had given this DROP) over 2 months ago

_TRUMP WAS TRULY ONE OF THE FIRST WORLD WIDE WHISTLEBLOWERS AND SAID CLIMATE CHANGE WAS A HOAX and he pulled all American U.S.. Dollars out of the Paris Agreement on climate change
_Trump led the way into revealing to the world that Mainstream media news was FAKE
_Now Trump is exposing the U.S. judicial corruption System and all the three branches of government.

The.] INFILTRATION [into the deep systems was masterfully maneuvered by military operations across the world.....

Have faith, Patriots .. Lots of good people in harm's way exposing a global, globalist deep State cabal regime.

⚠️ ⚠️ WARNING
_2024 the year ADRENOCHROME will be exposed on a massive level<
IT'S COMING!!!!!!!!!!!!

TRUMP'S CONGRESS SURE HAS A LOT OF CARDS ON THEIR TABLE.

Podesta. Wiener. Clinton. Rich. WikiLeaks.
Macfee , [EPSTEIN] , BIDEN LAPTOP
Obama, Gates, Microsoft, FBI cover-up. Cia operations.

It's all connected to a deep state MILITARY COUP. .. It's so connected to [DS] military dark ops. .... The Storm is a massive world operation////
And >]]]]] INFILTRATION [[[[[was the key

_
A Lot happening world Patriots as Cia EXPOSURE is happening with UFO. Human trafficking, military coup on Trump using social media. FBI. Using intelligence agencies.
_Congress leaders Exposing Fauci
The virus.
_Leaders going after Epstein report and list

(This is only the start 😉)
The world Elites hide their money. Vatican Recall all assets. Rothschilds hides money and goes private. Switzerland hides money in mountains.
_NATO/CIA failure to capture Russia (as Colonial Macgregor has described)
_Less 5% of Americans take new vaccines and boosters >

Q) The Storm Rider /Official Page:
_ the world health org stops the 2022 and 2023 PLANDEMIC (but they are trying again for 2024, safety measures are in place to expose the WEF. CDC. WHO, DAVOS, GATES, CIA REGIMEN.   And it's happening now and continues into FULL exposure of 2025
_leaders in several countries including U.S. senators and Congressional leaders start to expose a world globalist agenda connected to WEF who.

_
_So many things happened!!!!!!!

___
DOD spokesperson admits U.S. a broken corporation and military will step in at a certain time as they watch and monitor U.S. citizens waking up.

_
_Billions across the world reject vaccines <
and new boosters<

_
_hundreds and hundreds and hundreds of millions of people got RED- PILLED in 2023.

____
SUMMER 2024 WORLD REVOLUTION
AGAINST THE DEEP STATE.
THE FALL OF THE CABAL

The best is yet to come

The number # 1 country in the world with tier one databases and the best IT network is in New Zealand with the national healthcare, government and vaccine database . >And the one person in charge of building the data system and being in charge of the tier one network just came forward as a whistleblower with the hidden data from the New Zealand governments science, healthcare and data tier one records. And he shows that one in four people who took vaccines are dying.

Barry Young who was in charge of the Tier one days base says he felt it was his duty and his calling through God to do the right thing and expose the truth of the deaths from vaccines and he knew they were going to arrest him and bring him down for doing this in his interview to a top journalist in the country.

Within hours of the interview Barry Young was arrested and his lawyer was also being indicted along with the journalist (now the New Zealand government is trying to spin the story).....Fortunately the EXPOSURE and TRUTH is already hitting the world wide web and expected to hit over a billion views in a few days.

_The Pentagon who pushed vaccines now has their own doctor stepping forward and saying heart failure among the military has spiked over 900% increase in soldiers, this comes off their own data base of the military and was EXPOSED by Lieutenant Ted Macie.
>Pentagon data shows heart failure spiked nearly 1,000% among pilots in 2022: whistleblower

_Currently in Congress major Whistleblowers from the U.S. and top EU insurance agencies (who have massive data of insurance clients deaths happening directly after vaccine injections support Spiked by 700% , while the unvaccinated deaths stayed the same) are talking with Congress and getting their protection to come forward as whistleblowers.

_BEHIND THE SCENES]]; Congress and the Senate white hats are going after the vaccine cover-up and corruption, that will connect a worldwide ELITE operation being run by globalists (currently several senators and Congress have already divulged information on ELITES controlling WHO. connected to Gates > WEF)

> The Senate is waiting on military alliance operations in several countries to bring forward MAJOR Whistleblowers in healthcare industry, military networks and government orgs,>>> including the 2024 military leaks of the COVID creation and vaccine planned agenda dating back to 1999- 2007- 2013 <<<

_____

The MASSIVE walls are breaking< of the man made COVID pandemic, man made creation of the virus and man made death vaccines!!!
It's happening and growing in real time, and behind the scenes the healthcare industry and the doctors and their leaders are facing the truth that they helped kill millions of people around the world.....

The TRUTH is beginning to sink in and major panic is happening as healthcare systems around the world are asking the governments to censor the internet and shut down whistleblowers and free speech.
PANIC is happening as they know TOP WHISTLEBLOWERS ARE COMING FORWARD AND MASSIVE DATA LEAKS IS OCCURRING MYSTERIOUSLY<

_____

_The Globalist deep state cabal operations are failing on a grand scale.

BEHIND THE SCENES>];
MASSIVE PANIC INSIDE OF W.H.O. AND WEF AS THEIR COMPUTERS AND SYSTEMS HAVE BEEN CYBER-ATTACKED THE PAST 3 MONTHS
([They] try desperately to keep the data breaches hidden)

December 12- 2023 from Dr Mercola

Over the past year, the House Judiciary Subcommittee on Weaponization of the Federal Government has released several reports detailing how the government is

a) harassing and intimidating citizen to shut down undesirable viewpoints

b) using misinformation and propaganda to drive false narratives and

c) censoring protected speech using third parties

November 30, 2023, House Judiciary Committee held another hearing on the Weaponization of the Federal Government. Investigative Journalist Michael Shellenberger testified and sheared evidence about the existence of a group called the Cyber Threat intelligence League (CTIL), which consists of military contractors that are censoring Americans and using sophisticated psychological operations against us.

The CTIL was founded by a group of former Israeli and British intelligence agents who initially volunteered their cybersecurity services FOR FREE to billion-dollar hospital and health care organizations in the U.S.

CTIL also offers physical security and "cognitive" security, which are volunteered by U.S. and British military contractors.

The CTIL's plan to control the information landscape also includes using debanking of financial leverage, pressuring social media platforms to change their terms of service to facilitate censorship and platforming under the guise of " terms of services violations', and more.

⊸•⊰◆⊱•⊶

The Unseen Underworld: Unveiling the Sinister Network of 10,000 Deep Underground Military Bases Across the Globe

https://makegreatnow.com/the-unseen-underworld-unveiling-the-sinister-network-of-10000-deep-underground-military-bases-across-the-globe/

Beneath our very feet lies a dark and secret world, concealed from the prying eyes of the unsuspecting masses. The Unseen Underworld, comprising over 10,000 Deep Underground Military Bases (D.U.M.B.s), is a shadowy labyrinth that spans the globe. This clandestine network operates in the murky depths, shrouded in secrecy, as the ordinary citizen remains blissfully ignorant of the sinister reality that lurks beneath the surface.

The Global Conspiracy

Unraveling the layers of this diabolical plot reveals a staggering 10,000 D.U.M.B.s scattered across the world map, with the United States harboring a shocking 1,800 of these covert installations. What unfolds within these subterranean citadels goes beyond the realms of imagination, as tales of human captivity, innocent children held in the grip of darkness, insidious weapons, and bio-facilities paint a horrifying picture. The stuff of dystopian nightmares becomes a chilling reality within these hidden depths.

In the intricate web of these underground facilities, engineered super soldiers, borne from a nightmarish fusion of science and madness, stand guard. Their existence,

208

concealed in the shadows, is a testament to the mind-boggling complexity of the malevolent forces at play. Kill switches, treacherous traps, and dormant enigmatic elements create an atmosphere of perpetual danger, waiting to be uncovered by those brave enough to venture into the abyss.

Faith Amidst Darkness

Yet, in the face of this encroaching darkness, a glimmer of hope emerges. Faith becomes the driving force empowering humanity to confront the faceless, combat the unseen, and dismantle the fortress of manipulation that has entangled the world for far too long. This faith, a beacon of resilience, is the key to unlocking the secrets buried deep within the recesses of the Unseen Underworld.

The Stranglehold of the Super Elites

To understand the depth of the malevolent control that has woven its tendrils around the globe, we must confront the shocking statistics that govern our world. A mere 0.02% of the global population, the Super Elite, exercises unimaginable power over the destinies of 7.7 billion souls. This infinitesimal faction, along with the Elite and Million Elite, orchestrates a sinister symphony of control that extends its reach to the Global Military and Intelligence Complex, surpassing 800,000,000.

This spidery network infiltrates every facet of society, from corrupted police forces and news stations to management positions, militias, and mercenary organizations. Its influence extends to professions such as doctors, lawyers, scientists, and officials, permeating corporate behemoths like pharmaceutical companies and the insidious Big Tech industry.

With a web that ensnares over 190 nations worldwide, the challenge to untangle this web of control becomes a Herculean task.

Behind the scenes, puppeteers with names like the Khazarian bloodline, the Rothschilds, and the Rockefellers pull the strings of power, manipulating the masses and dictating global narratives. The Vatican, with its historical grip on 70 nations, and influential figures such as Obama, Gates, the Bushes, and the Chinese Communist Party (CCP), further complicate this macabre game of power.

Crafted meticulously over 1,400 years, this ensemble is designed to maintain the status quo of oppression. The stranglehold of these unseen overlords permeates history, shaping our reality and dictating the course of human events. The time has come to expose the puppeteers and dismantle their insidious control over the destiny of nations.

The Warriors in White Hats

Amidst the darkness, a glimmer of hope emerges in the form of the White Hats within the military. These valiant individuals tirelessly work to minimize casualties and restore justice, pushing back against the encroaching darkness that has plagued our world for far too long.
The era of complacency has come to an end, and humanity must brace itself for the revelations that are about to unfold.

The time for infighting, bickering, and petty disputes among ourselves has expired. We stand on the precipice of unimaginable events, where the events themselves hold more significance than the dates. A tide of revelation is poised to surge, washing away decades of deception and unveiling the unthinkable.

In the shadows of our daily lives, an unseen underworld festers, a testament to the elaborate web of control woven around us. Over 10,000 D.U.M.B.s exist across the globe, serving as reminders of the malevolent forces that manipulate our reality.

Yet, we are not powerless. Amidst the encroaching darkness, a glimmer of faith shines bright, emboldening humanity to confront the unseen, challenge the status quo, and reclaim our collective destiny.

The time for revelation is now. The age of ignorance is over. The stage is set for the unveiling of the unthinkable. Let us look beneath, beyond, and brace ourselves for what is to come. As we stand united against the malevolent control that binds us, we usher in an era where the truth will prevail, and the Unseen Underworld will be exposed for all to see.

LIST OF OVER 188 DEEP UNDERGROUND MILITARY BASES SITUATED UNDER MOST MAJOR CITIES, US AFBS, US NAVY BASES AND US ARMY BASES, UNDERNEATH FEMA MILITARY TRAINING CAMPS AND DHS CONTROL CENTERS

In America alone there are over 188 Deep Underground Military Bases situated under most major cities, US AFBs, US Navy Bases and US Army Bases, as well as underneath FEMA Military Training Camps and DHS control centers.

There are also many Deep Underground Military Bases under Canada.

Almost all of these bases are over 2 miles underground and have diameters ranging from 10 miles up to 30 miles across!

They have been building these bases day and night, unceasingly, since the 1940s. These bases are basically large cities underground connected by high-speed magneto-levity trains that have speeds up to 1500 MPH.

The Black Projects sidestep the authority of Congress, which as we know is illegal. There is much hard evidence out there. Many will react with fear, terror and paranoia, but you must snap out of it and wake up from the brainwashing your media pumps into your heads all day long.

David Wilcock December 19-2023

OPEN LETTER TO THE MILITARY

We the People are asking you to assist us to take our country and sovereignty back from the corporation and FED banksters and corrupt politicians. We are FIRING every last one of them!!! They no longer represent us; therefore, they are FIRED!!!! We are done being slaves and paying taxes that our gov wastes our $$ and it's not even backed by gold, which is fiat worthless paper by illegal and corrupt FED!!! We want the 2020 election fixed; we know you all know JB the Pedophile did NOT win. We have the right to dissolve the government when all other measures have been exhausted. We have tried January 6th protests, Convoy's, Multitudes of protests, going to school board meetings, Town meetings, Attending the Trump Rallies, the rightful winner!! Social Media Warriors Voting: election machines rigged & It was ALL rigged, See the movie RIGGED and Mike Lendel's Proof. We have lost friends, family members through covid, poisonous vaccines, and through sharing the horrific truth of all hidden from us for decades, Kennedy assassination and 9/11 is just two of 100s of examples we could give. We are tired of wars and paying for wars we DO NOT want!!! We do NOT want to support any wars unless enemies attack us. We want our country back and are demanding you act ASAP on our behalf. We want action, we want our sovereignty, we never agreed to be a mini corporation at birth and given all Capital letters on our birth certificates!!!! We never agreed to be traded like cattle while others make $$ from us unknowingly. We refuse to be slaves to the deep state for one moment longer. We want NESARA implemented ASAP. We will put another much smaller government in place once these horrible, inept people are fired and prosecuted for their crimes against us and our CHILDREN.

Patriots who agree please share. Military reading this please take this message from We the people to your commanding officers. Please respond to us using the EAS system. Patriots this needs to go viral on every platform. We the people have spoken. Military needs to act now on our behalf. Before we have no country left!!!!!

 Respectfully,

USA AMERICAN PEOPLE

(if you feel called shared this far and wide)

End Of 'The Year Message
Ashtar

End Of The Year Message – Ashtar – Rafael (Neva/Gabriel RL)

Revelations Of The Soul And Material Balance
Greetings, Family!
It's good to be here once again. Here I come with another message, my dear ones. It's good to receive affection from all of you.
It's good to know your support and that's what everyone needs: lovingly supporting each other and holding hands.

You are in the same boat of experiences and when you begin to see each other lovingly in this way, it will become easier for you to understand and accept each other as you are. This will boost your lives when you understand that you are not the enemy.

When you realize that together you become stronger. When you realize that you should never have been against each other, but all in connection with the Light for something greater.
That your union would take you to another state of consciousness and here you are, almost entering 2024, a year 8, a Saturnian year.

Each of you carries a package, some lighter, some heavier, others often almost unbearable to carry, I know. I know because I see and feel them, my dear ones.
You are in a great task and the fact that you chose this great mission to leverage your evolutions has greatly affected your soul core and now we are all in a healing work... of deep healing and alignment of each one of you.

Yes, we are here to help and, as I always say: do the work that fits each of you. We are your support and so you can continue to count on us, your Space Family.
My dear ones, you have immense support for this, more than you know. You have your Sun, your Astro-King. Many of your scientists are concerned about some things that have been happening there, with the masses of plasma being sent to Earth.

Let me tell you what it's about: it's about healing, it's about a plasma that will wash your souls. It is a bath of Light. The Creator does not play with lives. The Creator does not wish to see anyone suffering.

Creation is not disordered and nothing happens without a reason that leads to greater consciousness. You may not understand, in your human limitations, but your souls understand and are open to all the transformation that is coming, that is right before you. This is a long-awaited time, my dear! I've told you this before, haven't I?

2024: Saturnian Year, Year 8 Of Physical Balance And Conclusion

You are entering a special year where many of your material concerns will begin to be resolved.

Yes, my dears, but before you think about the money in your accounts, I want you to understand this as balancing, first and foremost, your physical bodies.

The souls of your organs are being strengthened. Your physical body is being strengthened and this is what the solar projections have caused.

— Will there be some interference to your satellites due to these projections? The answer is yes, but don't be scared by this.

— Will there be an increase in volcanic activities? The answer is yes. Don't be scared by this.

— Will there be disturbances in the Earth's core that will cause some tectonic adjustments? The answer is yes, but don't be scared by this.

— Why don't you need to be scared? Because you are you – as I always say – and you came to Earth for this: to help in this entire healing process that Earth is going through.

2024 is a year of physical balance, both for you, my dear ones, and for Mother Earth herself.

Your material life will begin to come into harmony because you will physically be in harmony. Your bodies need healing.

How many years have you suffered from chronic pain, without your doctors being able to understand what the cause is? When no exam reveals the reason?

Yes, my dear... ENERGY. I'm talking about ENERGY – which needs to continue its flow. And when this flow is interrupted it causes physical pain, and that is what you will begin to understand perfectly.

Many of you already understand this, theoretically speaking, but you will see this in practice in 2024. Oh, summer!... I know you will because I'm in 2024 and I already see it happening. There's no escaping the progress, dear ones! Everything moves forward>

The Creator has plans for this planet, very beautiful plans. The Creator has looked at each of you with so, so much love that there is nothing that the evil of this world can do to disrupt HIS plans for you. Do not fear evil. Do not fear the traps that this evil can try against you.

I, Ashtar, am in charge of taking care of this evil, personally, and I can say: I have never failed to fulfill a mission given to me and I was given the mission to nullify the action of evil on this planet... AND I WILL FULFILL IT. I GIVE YOU MY WORD.

I'm not alone here. There is an army of Space Beings with me; an army larger than a thousand populations on Earth combined. It's more than you can imagine. Can you understand this? There is nothing to fear, my dear ones! 2024 brings the astrological force of Saturn, my dears: the ruler of discipline and divine lessons.

It's a year where you will have to strengthen your foundations, because everything that is karmically stagnant will come to your healing. That's good, isn't it?

Release what needs to be released, definitively heal what needs to be healed.
Saturn will teach you valuable and uplifting lessons through challenges that will require patience and continued determination from you.
It is time to face... "head on", the responsibilities that fall to each of you, with wisdom, to stop outsourcing what is yours.
You will receive strength. You will notice it emerging from within you.
This is also a year where you will begin to see who you really are, because what many use to maintain themselves in some environment can collapse, as it is not the real experience of who you certainly are! Yes... if you wear a mask to maintain yourself in any environment, this will fall apart, because 2024 demands authenticity.
It demands that you be you in your real essence and Saturn will call for this, continuously, in the 12 months ruled by it.

This astrological force will demand that you commit to the truth of your souls, your real purpose and, of course, to resolve what needs to be resolved, without delay, simply because there is no more time to waste!
Take advantage of this Year 8, dear ones, when you will be focused on material stability, high planning capacity and a great phase for new projects and inner priorities, returning to the INFINITE capabilities of your SOUL.

Self Esteem

My dear ones, your self-esteem has been shaken since your arrival into this world. You have lost your confidence in yourself. You have seen the love in your hearts grow cold; you were lost in wars; you forgot yourselves and begged for love throughout this planet; you humble yourselves; You gave your lives into the hands of others and, mutually, you sought love from each other, but not in a balanced and unconditional way.

You looked for someone who could fill the voids you carried; you were looking for something that would connect you again to the Primordial Source of true Love.

It was eons of time in the darkness of a disconnection that made you believe that you were not worthy of Love, that you were not good enough, that you were not loved and cherished.
Yes, my dear! I've been following you since the beginning and I know what I'm talking about… because I see it.

I have seen every moment of extreme despair; I also saw moments of hope and confidence that everything would get better.
I saw you become discouraged again and regain confidence. I followed and followed this rise and fall of vibrations inherent to this dual dimension.
My dear ones, solar projections will make you look at yourself and connect with this Source of Love within you. The time has come for that.

You will realize that this inexhaustible source of love gushes from your heart. It was always like this, but you forgot about it.

You will see this in practice in 2024, because this is also a year of regaining your self-esteem.

Revelation Of 'The Soul

Yes, my dear! There is a connection and harmonization with the three kingdoms of your nature.

Your physical body is harmonizing with your soul, your soul harmonizing with your spirit and this loving fusion reconnects to Source and the Creator.
This is what this year will bring for you, if you believe and allow yourself to live this process.
Stay calm: everything is under control. Your soul is revealing to you the new that, deep down, is not new, but that was hidden by the pain of life in the dimension you are in.
What your soul wants to reveal to you in 2024 is more than anything you have ever experienced in your entire current life and even in previous lives.
Can you believe this? And much more!
Your soul will lead you to a Sacred Space within yourself that you have not accessed for a long time.

And I'm not talking about the etheric personality soul…, I'm talking about your Higher Self, your Monadic Soul. I'm talking about your SOUL in loud letters.
It will lead you to reintegration with all your bodies and the beginning of the fusion of your four lower bodies with your HIGHER ones.

You have waited for this for a long time and you will see it happen in this lifetime, do you believe it? Same? So so be it!
What else can your SOUL reveal to you, besides what I have already said? Your dormant gifts?

Your ability to connect with beings from other dimensions more easily? The ability to hear GOD like never before? Precise intuition? Deep empathy capable of understanding what the other feels like never before and enabling genuine forgiveness?
These are JUST SOME THINGS that are coming, that Solar Plasma is bringing.

Obviously, many still asleep will see these Revelations of their SOUL as a threat. They will see Solar Plasma as a risk of extinction for the Gaian human species.
Yes, many still don't understand like you do. Be calm with them; they will also arrive at the point of SOUL REVELATIONS.
No one will be left without help. Take it easy with them. This is what you were sent to do. To help at this time. There you are.

'The 'Three Chakra

My dear ones, all of this will generate immense activity in the three chakras of the seven that you know most: your basic chakra, your sacral chakra known as sexual and your umbilical chakra known as solar chakra.

All activity will be there in 2024. You will go through a great alignment at these points. Solar projections will intensify this. Remember to maintain a good diet, good hydration and physical activity.
This will be essential for you in 2024. Your physical bodies will receive great forces coming from the Sun and you, as their ruling consciousness, need to collaborate by giving them what is necessary: good care.

Do not forget that your bodies are blessed vehicles through which your consciousness can operate in this hologram, at least until you relearn the mastery of physics and its realms.
Oh, my dear! There are so many more things I could tell you about what I see in your future!
These are times of a lot of work and also many good revelations!

Nothing like time to show that it was all worth it! Nothing like time to heal the wounds that you carry, from such difficult experiences, but which were enormously uplifting!

I See you here and now, reading my words!
How many people in the world are allowing themselves a moment like this, of connecting with a Message from Space?
Even more so, who would believe that there is a Star Being speaking through a mediumistic device, the channel transmitting my message, now?

Yes, you believe it because you know it to be true! And not just because you know the nature of the channel that transmits my message, but because you feel it.

It takes more than simply believing; you have to feel it... and you do>
You know it's me, Ashtar, speaking here, and it's not your minds and logic that reveal this: it's the SOUL. YOUR SOUL REVEALS THIS.

My beloved, beloved Earth Family, this is enough for now. That's enough. We'll see you more often. Our intensive connections are just beginning.

I love you. You are My Family!

And so it is.
Blessings!
Your brother,

Ashtar.

Never Be Afraid, Jus1 Be Prepared

– Kabamur – Pleiadian Collective

The best really is yet to come. These aren't just words. I am here to tell you that there is an unstoppable plan in motion. I'm talking about something bigger than New Financial Systems or Mass Arrests.

We are soon to experience 'heaven on Earth' with beauty beyond imagination.
I believe many truth seekers are unprepared for what's coming before the Shift because they're being told white hats are in control of world events and are about to rescue us from evil.

This is dangerous thinking. This is the great tribulation and we're not at the end yet. There will be a continuation of what we're already seeing, which will escalate drastically by the next election.

Freedoms will continue to be taken away, brainwashing of children will continue, globalist agendas will expand, more clones, trans-humanism, bank collapse, climate blackouts, power grid attacks, food shortages.

The Galactic Federation won't stop every bad thing from happening but we are being continually assisted in ways we can't see.

They will weaken pandemics, they will stop nuclear weapons and they won't allow any fake alien invasions.

It's okay to be aware of dark plans but don't dwell in negative thinking. This lowers your frequency and allows for astral attachments.

Be mindful of your thoughts and actions because this creates your experience.

Keep your Light. Don't lose hope. We are not prisoners and everything is temporary. Continue standing for truth but know what is within your control to change and have peace about what you can't change.

Focus on the big picture and ignore the daily noise.
Go within to connect with your Higher Self to find peace.

Never be afraid, just be prepared.

WE Love you so!

ACKNOWLEDGEMENTS

My thanks and appreciation to my teachers, who are many. Thanks to the authors of all the many books whose wisdom I have absorbed by reading, studying and assimilating their words. They have opened me up to see life in a larger way and have helped give me the ability to apply the law of opposites over the years. The time of duality is over, we need to learn to think in a different way now.

And thanks to all the others who have come since, in person and online, with their wisdom and guidance and research of the truth. All of you have helped me discover that part of my life's purpose which gave me the power and strength to write this book.

About the Author

Rosanna Martella is an amazing artist living in the Northeast USA. As a sculptor and painter she is very active and was surprised when she found out she had epilepsy. She healed herself from this incurable condition, naturally and wholly. She published her book *Healing Epilepsy Naturally a True Story* where she explained how she healed herself.

Rosanna Martella is a counselor and consultant who assists others in their journey to regain health. She has cured many people around the world. Let her wisdom and knowledge be your ticket to healing and peacefulness.

In her entire life she never thought about writing this book, but she got tired of hearing lies and deception, by satanic evil ones that have used bioweapons disguised as medicine they called vaccines that killed millions of humans. She pushed herself to write this book.

The Truth Must Prevail.

If you choose to work with me in a
Consulting/Coaching Program for an entire
year, focusing on detoxing and getting healthier,
How much better could you feel
over the next 12 months?
1-856-782-7310 (North America).
macro@rosanna.com

EMAIL macro@rosanna.com

WEB SITE www.thehealthmode.com

www.ingramcontent.com/pod-product-compliance
Lightning Source LLC
Chambersburg PA
CBHW081811200326
41597CB00023B/4219